PERGAMON INTERNATIONAL LIBRARY
of Science, Technology, Engineering and Social Studies
The 1000-volume original paperback library in aid of education, industrial training and the enjoyment of leisure.
Publisher: Robert Maxwell, M.C.

PHOTOGEOLOGY AND REGIONAL MAPPING

Other titles of interest:

ANDERSON: The Structure of Western Europe

ANDERSON & OWEN: The Structure of the British Isles

BOWEN: Quaternary Geology

CONDIE: Plate Tectonics and Crustal Evolution

*MELCHIOR: The Tides of the Planet Earth

OWEN: The Geological Evolution of the British Isles

SIMPSON: Geological Maps

*Not available on inspection

PHOTOGEOLOGY AND REGIONAL MAPPING

by

J. A. E. ALLUM

B.Sc., Ph.D., P.Eng., C.Eng., A.R.S.M., F.I.M.M.

Inco Metals Company

PERGAMON PRESS

OXFORD · NEW YORK · TORONTO

SYDNEY · PARIS · FRANKFURT

U.K.	Pergamon Press Ltd., Headington Hill Hall, Oxford OX3 0BW, England
U.S.A.	Pergamon Press Inc., Maxwell House, Fairview Park, Elmsford, New York 10523, U.S.A.
CANADA	Pergamon of Canada Ltd., 75 The East Mall, Toronto, Ontario, Canada
AUSTRALIA	Pergamon Press (Aust.) Pty. Ltd., 19a Boundary Street, Rushcutters Bay, N.S.W. 2011, Australia
FRANCE	Pergamon Press SARL, 24 rue des Ecoles, 75240 Paris, Cedex 05, France
FEDERAL REPUBLIC OF GERMANY	Pergamon Press GmbH, 6242 Kronberg-Taunus, Pferdstrasse 1, Federal Republic of Germany

First edition 1966
Reprinted 1969, 1975
Reprinted (with corrections and additions) 1978

Library of Congress Catalog Card No. 66-23842

Printed in Great Britain by Biddles Ltd., Guildford, Surrey.

ISBN 0 08 012032 6 flexicover
0 08 012033 4 hardcover

Contents

List of Plates

The following are Royal Air Force photographs, Crown Copyright reserved: Plates IIIB, IVA and B, V, VIA and B, VII, VIII, X, XI, XIIA, XIIIA and B, XIVB, XVA and B, XVIA and B.

Preface

PHOTOGEOLOGY is the name given to the use of aerial photographs in geological studies. Aerial photographs are a source of geological information that may be unobtainable elsewhere. If the fullest use is to be made of this information for the purpose of geological mapping, there must be a planned integration of the photogeological work with the field and laboratory investigations. This book outlines a method by which this may be achieved.

It includes, amongst other things, a discussion of: (a) the geological interpretation of aerial photographs; (b) the compilation of these interpretations on to maps; (c) the use of aerial photographs in the field; and (d) the use of aerial photographs for the production of the final geological map. I hope that it will be of value to all those concerned with the earth sciences whose training includes an elementary knowledge of geology.

The term "regional mapping", as used in this book, means the mapping of large areas. The use of aerial photographs in the field and for compilation purposes is discussed from the point of view of a geologist responsible for mapping a large area in a short time. I believe that many of the suggestions made, particularly those concerning field work, will also be of value to geologists mapping in a more detailed manner.

I have not hesitated to include personal opinions that may require qualifying in the future. It is seldom possible to prove conclusively a photogeological point by a limited number of correlations between stereopairs and the corresponding field outcrops. If photogeology is to advance, geologists must be prepared to publish opinions and make generalizations based on their own experience with the photographs in the field. Other geologists may then amend the generalizations as a result of further field experience with photographs in different geological environments. Thus, progress will be made.

Acknowledgements

MY THANKS are due to: the British Government for permission to publish the photographs of which they hold the copyright; the Governments of Bechuanaland, British Guiana, Kenya, Nigeria, Somali Republic, United Republic of Tanzania, and Zambia for permission to publish the photographs of their countries; C. F. Casella & Co. Ltd., Wild Heerbrugg Ltd., and Thorn Electrical Industries Ltd. for supplying the photographs reproduced in Plates IA and B, IIA, and IIIA respectively; the Ministry of Overseas Development and the Overseas Geological Surveys for permission to publish the book; Mr. J. A. Eden of the Directorate of Overseas Surveys, and Mr. J. E. G. W. Greenwood, Dr. J. V. Hepworth, Mr. E. A. Stephens, and Mr. G. Whittle of the Overseas Geological Surveys, Photogeological Division, for reading one or more of the various chapters; Mr. W. A. Cook, of the Overseas Geological Surveys Library, for his help over the years in obtaining innumerable photogeological references; and my wife for typing the various drafts.

The description of the minor control plot (Chapter 10) has been influenced by that in the *Admiralty Manual of Hydrographic Surveying* (1948), and that of stereoscopic vision by *The Psychology of Perception* (Vernon, 1962). Parts of Chapters 4, 5, and 6 are based on the paper published by the Institution of Mining and Metallurgy (Allum, 1961).

The use made of the *Manual of Photographic Interpretation* (American Society of Photogrammetry, 1960) for definitions, and the published records and maps of the geological surveys of the countries mentioned above, is acknowledged. All errors and omissions are, of course, my fault.

J. A. E. ALLUM

CHAPTER 1

The Aerial Photograph

THROUGHOUT this book, the expression "aerial photograph" refers to a photograph taken from the air with the camera pointing vertically downwards at the time of exposure. There are other types of aerial photographs, such as the **high oblique,** in which the camera is tilted sufficiently to include the horizon, and the **low oblique,** in which the camera angle of tilt is insufficient to include the horizon. The great majority of all aerial photographs used for mapping purposes are of the vertical type. A typical aerial photograph paper print with added titling strip is illustrated in Fig. 1.

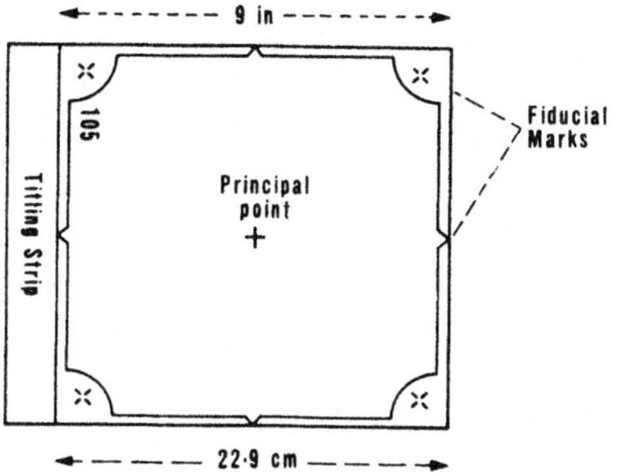

FIG. 1.

Many photogrammetric definitions sound complicated. The **principal point** is defined as the foot of the perpendicular from the interior perspective centre to the plane of the photograph, i.e. the foot of the photograph perpendicular (American Society of Photogrammetry, 1960*). For our purpose, it is sufficient to regard the principal point merely as the geometrical centre of the aerial photograph; it is, in fact, sometimes referred to as the centre point.

* References are given in full in the References section which appears on p. 104.

The **fiducial marks,** occasionally referred to as **collimating marks,** are index marks rigidly connected with the camera lens through the camera body, which form images on the negative (op. cit.). One of their purposes is to define the position of the principal point; if two straight lines are drawn between opposite fiducial marks on the photographs, they intersect at the principal point.

The fiducial marks can be of various designs; they are located either in the corners or along the edges of the photographs.

Each photograph should have recorded on it: flight number, part number if any, flying height of plane, focal length of camera lens, date of photography, and country represented. This information may be recorded on a titling strip, as illustrated in Fig. 1, or on a photograph of an instrument panel reproduced at the side of the print.

HOW AERIAL PHOTOGRAPHS ARE TAKEN

When the same piece of land is photographed from two different positions in the air and the resulting two photographs inspected simultaneously with a stereoscope, an apparently solid model of the ground is seen. This is known as the **stereomodel.** The three-dimensional stereomodel is so valuable for mapping purposes that virtually all aerial photographs are taken in such a way as to provide it.

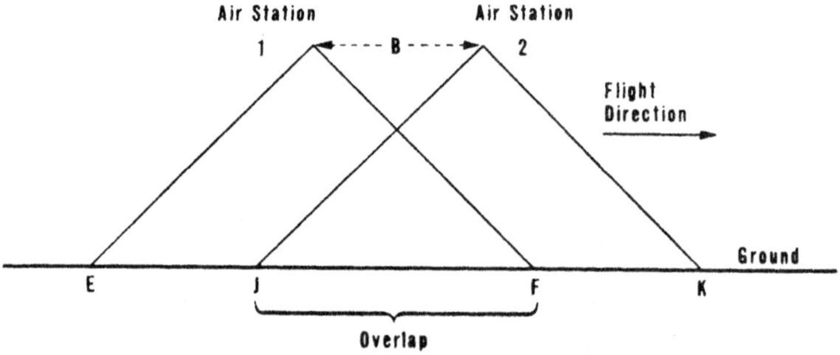

Fig. 2.

Consider Fig. 2, which represents a vertical section in the line of flight of the plane. A photograph taken from Air Station 1 covers the ground *E–F*. The plane flies to Air Station 2, and photograph 2 is taken, which covers the ground *J–K*. The ground *J–F*, known as the **overlap,** is thus represented on both photographs.

The amount of overlap depends upon, among other factors, the **air base** B, which is the distance between the air stations. If the overlap were 50 per cent, *J–F* would be half *J–K*, and when photograph 3 was taken from Air

Station 3, the new overlap would be *F–K*; thus the whole of the area covered by photograph 2 would appear on two photographs and could be viewed stereoscopically using first photographs 1 and 2 and then 2 and 3. If the overlap were less than 50 per cent there would be a gap in the centre of photograph 2, which was not included in either overlap, and which could not be viewed stereoscopically; to avoid this, an overlap of 60 per cent is usually aimed at.

If the overlap is 60 per cent, every point on the ground is represented on at least two consecutive photographs; it is therefore possible to divide the photography into two sets of photographs, each of which covers the whole

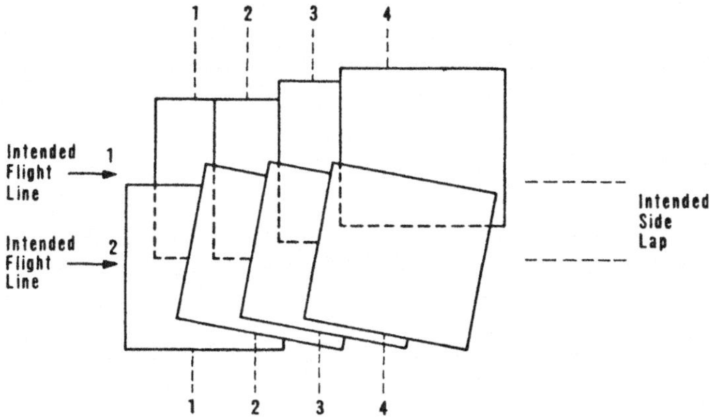

Fig. 3.

area; one set will consist of photographs 1, 3, 5, etc., and the other of photographs 2, 4, 6, etc. Either set can be referred to as a set of **alternate photographs;** the other set is then referred to as the **conjugate photographs.**

To photograph a large block of ground, it is necessary to fly a number of parallel strips; these must overlap laterally (see Fig. 3) to ensure that no area between them is left unphotographed. This lateral overlap, known as **side-lap,** is usually about 30 per cent. If the terrain is flat, a smaller side-lap of (say) 20 per cent would be suitable; but if the terrain is mountainous, and particularly if the photographs are likely to be required for mosaicing, a larger side-lap would be desirable (see Corten, 1962, and Kilford, 1963).

ERRORS IN FLYING

If the plane is affected by a cross wind and no correction is made, a condition known as **drift** occurs; the photograph edges remain parallel to the intended flight line, but the plane itself drifts further and further from that line. This is illustrated by photographs 3 and 4, of Flight Line 1, in Fig. 3.

If the plane is corrected for drift by turning into the wind, without the camera being reorientated, a condition known as **crab** results; the photograph edges cease to be parallel to the intended flight line, though the movement of the plane may follow the line closely. The result is that the side-lap is irregular and not as much as intended. This is illustrated in photographs 2, 3 and 4, of Flight Line 2, in Fig. 3. Thus the correct and usual practice is to reorientate the camera in the plane, when the plane turns into the wind to correct for drift; in this way, the errors of both drift and crab are obviated.

If the plane is not horizontal at the time of exposure, the resulting photograph is said to be tilted. This is a very obnoxious error; it makes stereoscopic viewing uncomfortable for the eyes, and produces slopes in the stereomodel that are unrelated to reality; in extreme cases, it may make rivers appear to flow uphill.

If the plane speed is not correctly related to the time interval between exposures, the overlaps will be either too great or too small; if the overlaps are too great, the stereomodels will show insufficient relief; if they are too small, there will be strips across the photographs that are incapable of being viewed stereoscopically.

VARIOUS USES OF AERIAL PHOTOGRAPHS

The property of aerial photographs of providing a three-dimensional, permanent, overall view of the ground, at almost any scale demanded by the user, makes them of great value for such activities as topographic mapping, regional geological mapping, regional soil mapping, forestry, land-use studies, military intelligence, archaeology, and civil engineering (see American Society of Photogrammetry, 1960; Avery, 1977; Lueder, 1959; Newton, 1971; Verstappen, 1977).

SPECIAL TYPES OF AERIAL PHOTOGRAPHS

Unfortunately, the many different users of aerial photographs do not all require the same type of photography. The civil engineer, who perhaps needs to contour a reservoir site very precisely, may require large-scale photography at an average scale of (say) 1: 5000; the geologist, however, making a rapid reconnaissance geological map of a large area, may prefer smaller-scale photography of (say) 1: 60,000. Foresters find they can distinguish hard-wood from soft-wood trees more easily if they have infra-red photography. Archaeologists require photography taken at a particular time of the day and year, in order that the foundations of ancient sites should be revealed in the crop markings. It is impossible for any one type of photography to satisfy completely all these different users.

If the importance of the project justifies it and if the money is available, special photography may be taken. Such conditions seldom apply to geologists mapping on a regional scale however.

Most photography, except that for special, local projects, is taken for topographic mapping purposes according to specifications laid down by the Survey Department of the country concerned. It is usually taken on panchromatic film at a scale between 1: 20,000 and 1: 50,000. This is the photography that is normally available to the geologist, forester, and land-use scientist. Geologists have got so used to it that they seem to have stopped considering what the ideal photography for their purposes is; they tend to like what they are familiar with, and in any case, have to accept what they are given. This question, as to the best photography for geological purposes, would form an interesting and valuable subject for research; it has been so neglected that it is impossible to give an authoritative answer (Corten, 1962).

It is known that with large-scale prints (1: 10,000 or larger), photogeological interpretation can be very difficult. Of course, prolonged experience in the field with such prints might enable the geologist to develop new interpretative criteria, which would make them valuable for photogeological purposes; but this is not certain. It is also known that at very small scales (1: 80,000 or smaller), annotation of the prints becomes difficult and many structures are overlooked altogether. The optimum scales for photogeological work may therefore be about 1: 20,000 for detailed work, 1: 40,000 for regional work, and between 1: 50,000 and 1: 80,000 for rapid reconnaissance work. These figures refer to the scale of the paper prints on which the work is done, and not to the original negatives. Much more use could be made of the ability to print positives at different scales from the same negative. If the photography (negatives) is at a scale of (say) 1: 40,000, prints could be supplied at both 1: 20,000 and 1: 60,000 scales. This could, for some areas, be a great assistance to the geologist.

Attempts are being made to determine the value of colour photography for photogeological work (American Society of Photogrammetry, 1968; Allum, 1970; Norman *et al.* 1971; Fischer, 1958). There seems little doubt that the extra criterion of colour is of value in photogeological interpretation. It is unlikely, however, that colour photography will be widely used for regional geological mapping until inexpensive, good quality, colour paper prints become available, at the suitable scales mentioned above. Colour transparencies are inconvenient for use in the field.

ORDERING OF AERIAL PHOTOGRAPHS

The method of obtaining aerial photographs varies from country to country. The Survey Department of the relevant country is the most likely source of reliable advice because, in most countries, it is the largest user of aerial photographs (see also American Society of Photogrammetry, 1960). In Britain, the organization to consult is the Ministry of Housing and Local Government.

HANDLING OF AERIAL PHOTOGRAPHS

The mapping of a single 30-minute sheet may require the use of several hundred photographs. Some definite procedure for organizing and handling the photographs must therefore be developed and adhered to.

There are two problems: firstly, to determine which photographs are required for different parts of the area; and secondly, to select quickly, from the large collection, the particular photographs required.

The first of these problems is solved by the use of either a cover diagram or print laydown. A **cover diagram** merely shows the approximate latitude and longitude of the area covered by each strip of photography; the photograph numbers are recorded at either end of each strip on the diagram, so that it is possible to estimate the approximate serial number of the photograph corresponding to any point within the strip. Cover diagrams are usually held by the agency responsible for the negatives; tracings or photographic copies of them can easily be made.

A **print laydown** consists of a single photograph of the alternate prints of the original photographs, which have been laid down in approximately their correct relative positions. A very similar compilation is the **photo index,** in which all the prints are laid down before being photographed. The photograph serial numbers are printed at the end of each strip of photography on the print laydown, and on each photograph of the photo index.

Print laydowns or photo indexes of an area enable the serial number of the photograph corresponding to any particular part of the area to be determined instantaneously; they are therefore much more useful than cover diagrams.

In order to solve the second problem, it is necessary to file the photographs according to a definite system. If all the photographs for the area are numbered in a consistent manner, it is possible, and very convenient, to store them in a pile in strictly numerical order. For instance, if all the photographs are marked with sortie, flight, and serial numbers, they are obviously amenable to filing in strict numerical order. If the sortie number is 72A, the flight number 45, and the photograph serial number 5027, the photograph is completely specified by the number 72A/45/5027. All photographs numbered in this manner can obviously be kept in strict numerical order.

Commercial aerial photography might well be numbered according to country of origin, flight number, and photograph serial number. Suppose the country was U.R. Tanzania (TAN), the flight number 22, and the photograph serial number 036, the photograph would be completely specified by the number TAN/22/036. With this manner of numbering also, the photographs can be conveniently kept in strict numerical order. Even if several countries are involved, the system is still satisfactory, providing that attention is paid to the initial letters (signifying the country) and that the photographs are kept in strict alphabetical and numerical order.

The system can be difficult to use if the cover includes photographs numbered in different ways. This is found when photography of different ages and contracts has to be used for the same area. The solution to the problem in these circumstances is to: (a) divide the area into photograph strips; (b) number each photograph strip; and (c) select all the photographs for each strip and put them into an envelope, marked with the strip number. Print laydowns are particularly useful for this latter system because they indicate the photograph serial numbers, and also because they can themselves be annotated with the strip numbers.

PRINT LAYDOWNS AND MOSAICS

Print laydowns are of great value to geologists because they: (a) give him a "bird's eye" view of the area; (b) indicate areas of particular interest; (c) show the access routes to these special areas; (d) indicate the most probable positions of outcrops; (e) simplify the planning of traverses; (f) indicate the serial numbers of the photographs required for a particular traverse, or for any particular area under discussion; (g) simplify the selection of camp sites; and (h) provide illustrative material for scientific papers, and committee discussions. In the absence of controlled topographical maps, they can be used as uncontrolled base maps for the compilation of reconnaissance field mapping. In short, they are of value at nearly every stage of major geological projects.

The only geological mapping one can envisage in which print laydowns might not be of value, is that of highly detailed work in small areas. Even under these conditions, the geologist might welcome the regional view of his area that the print laydown would give him.

The geologist may make his own print laydown as follows:

(1) Separate the alternate photographs from their conjugate pairs and inspect them to ensure that they provide 100 per cent cover of the area. (In areas of insufficient photographic overlap, it might be necessary to use a few of the conjugate photographs.)
(2) Remove all the edges of the alternate photographs except the titling strips; this can be done with scissors, but is easier with a guillotine.

(3) Select a strip of photography near the centre of the area, and join the adjacent alternate photographs to one another with cellulose tape in their correct relative position; this is done merely by superimposing corresponding points of photographic detail. The photographs should be overlapped so as to hide the titling strip.

(4) Select the alternate photographs of the strip immediately to the south of the strip just completed (3) (for the purpose of this description it is assumed that the strips lie east–west). Attach these photographs to the completed strip, one at a time, from west to east. This stage is less easy than (3) because it is necessary to attempt to make the edge photographic detail of each photograph coincide with that of the first strip (3) and also with that of the previous photograph of the new strip. In general, because of scale changes, perfect coincidence of both is not achievable; it is essential, however, that coincidence of photographic detail along the southern edge of the second strip should be achieved, because otherwise it will become progressively more difficult to add subsequent strips. The south-west corner of each new photograph is therefore put in coincidence with the south-east corner of the previous photograph and attached with cellulose tape. The new photograph is then revolved around its south-west corner into the position of best agreement with the photographic detail of the first strip (3), and fixed with cellulose tape. The second strip is completed in this manner.

(5) Third and subsequent strips to the south are added in the manner described under (4).

(6) The whole process described under (4) and (5) is repeated for all the strips to the north of the first strip.

The geologist making his own print laydown at original photograph scale need not proceed beyond stage (6).

The Directorate of Overseas Surveys (D.O.S.) make many print laydowns as soon as the relevant photography is received, and before maps are available to provide control. They proceed in much the manner described above, except that the photographs are fixed with mapping pins to a board, in their correct relative positions. After stage (6), they continue as follows:

(7) The names of principal towns, villages, and rivers are printed on the laydown.

(8) A border is added on which is printed the name of the country depicted, the name and number of the print laydown, the approximate latitude and longitude of the area covered, the serial numbers of

the photographs at the end of each strip, and a scale line representing approximately 10 miles.

(9) The print laydown is then photographed at a scale of about 1: 125,000. From the resulting negative, any number of print laydowns can be obtained at different scales, as required.

The D.O.S. has made many hundreds of print laydowns, mostly of areas in the less developed countries of the Commonwealth. Copies of these laydowns or their negatives have been sent to the relevant overseas Survey Departments and thus are available both in Britain and in the overseas countries concerned.

If there is no objection to cutting the prints and rendering them unserviceable for any other purpose, a more attractive form of print laydown can be made, which is known as an uncontrolled **mosaic.**

To make such a mosaic, the procedure is much the same as for the print laydown described above. The differences are firstly, that the photographs are not attached to one another but are stuck down, one at a time, on a compilation board with rubber solution, and secondly, that the corners and the edges of the photographs are cut away with a razor blade before sticking down, so that only the central portion of each photograph is used. With this method, it is possible to select the line of join between adjacent prints so that the mis-match of tonal values and the apparent mis-match of topographic features are reduced to a minimum. The production of mosaics is well described by Department of the Army, 1955.

The advantage the mosaic has over a print laydown is that it looks more like one continuous photograph and less like a patchwork of photographs.

Uncontrolled mosaics and print laydowns share certain undesirable characteristics about which the potential user should be warned. The scale varies all over the compilation (the cause of this is discussed in Chapter 2). Because of the variation in scale, occasional losses, duplications, or breaks in topographic details are apparent at the joins of photographs.

These undesirable characteristics can be reduced by making a controlled mosaic. Controlled mosaics of the highest quality require ground control and preferably scale-corrected, rectified, matched prints. (Rectified prints are those which have been corrected for tilt, and matched prints are those which have been hand printed to have corresponding tonal values.) The production of such a mosaic may prove an expensive, time-consuming task.

CHAPTER 2

Some Geometrical Properties of Aerial Photographs

SCALE

The scale of a photograph depends upon the flying height of the aeroplane above the ground and the focal length of the camera lens. A vertical section is illustrated in Fig. 4.

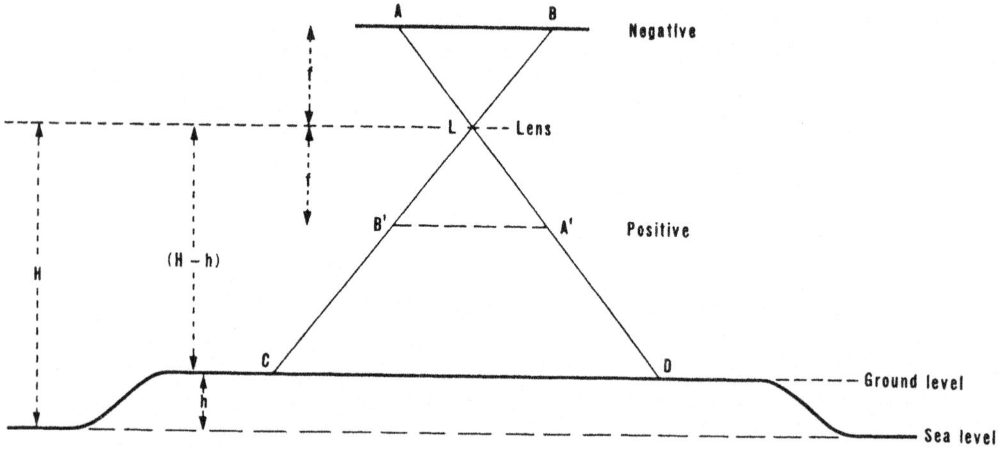

FIG. 4.

The height of the aeroplane above sea level is H and that of the ground is h; thus the height of the aeroplane above the ground is $H - h$; L is the camera lens of focal length f.

In reality, the camera takes a photograph negative A–B, which is then processed to produce a positive print, represented in Fig. 4 as A'–B'. Henceforth this subtlety will be ignored, and it will be assumed that the camera takes a photograph positive in position A–B. This simplifies the diagrams and does not affect the discussion.

The distance *C–D* on the ground is represented by the distance *A–B* on the photograph. The **scale** *S* of the photograph, sometimes called the **representative fraction,** is therefore the distance *A–B* divided by distance *C–D*, which, by similar triangles, is $\frac{f}{H-h}$.

Then
$$S = \frac{f}{H-h}. \quad (1)$$

It is important to remember that in this formula (1) f, H, and h must be in the same units. If H and h are in feet, and f is recorded as 6 in., then f must be converted into feet for use in the formula.

Suppose that in Fig. 4:

$$H = 16{,}000 \text{ ft}$$
$$h = 1000 \text{ ft}$$
$$f = 6 \text{ in.} = \tfrac{1}{2} \text{ ft.}$$

Then
$$S = \frac{\frac{1}{2}}{16{,}000 - 1000} = \frac{\frac{1}{2}}{15{,}000} = \frac{1}{30{,}000}.$$

Then the photograph scale $= S = 1:30{,}000$.

In Fig. 4, *C–D* is drawn horizontal, but a normal ground surface undulates. Imagine a 100 ft cube on *C–D*; the new h, at the top of the cube, would be 1100 ft, and thus the scale *S* at which the top of the cube would be recorded on the photograph, would be 1: 29,800. Similarly, if a 100 ft cube were dug out of the ground *C–D*, the h for the bottom of the cube would be 900 ft, and the scale *S* for the bottom of the cube would be 1: 30,200. Thus the scale of a photograph of undulating country varies all over its surface, with the lower points being depicted at a scale smaller than the average, and the higher points at a scale higher than average. The effect is rather as though a planimetric map were printed on a rubber sheet, and then the areas representing low ground were compressed and those representing high ground stretched.

The term "photograph scale", therefore, does not mean the scale at a particular point, but rather the average scale over the whole photograph.

It is worth remembering that, in the discussion of maps and photographs, a small scale means a small-sized, and a large scale means a large-sized, representation of the area.

It will now be obvious that, if the aeroplane flies at a constant height above sea level, whilst it takes many photographs, the average scale of the photographs will vary as the average elevation of the ground changes. If, therefore, photography of constant average scale is required, then it is necessary for the aeroplane to fly a constant height above the average ground elevation.

PARALLAX

The apparent displacement of the position of a body with respect to a reference point or system caused by a shift in the point of observation, is known as **parallax.** With aerial photographs, it is the point at which the photographs are taken which changes. In photogeology, therefore, the parallax of a point is the apparent difference in position of the point on two consecutive photographs. For photogrammetric purposes, a more rigorous definition is necessary, but this is sufficient for us (American Society of Photogrammetry, 1966).

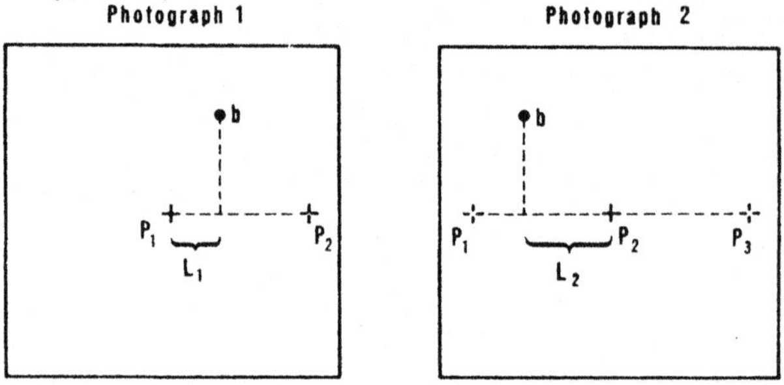

FIG. 5.

In Fig. 5, P_1, P_2, and P_3 are the principal points of the first three photographs of a strip. Because of the 60 per cent overlap referred to in Chapter 1, the point on the ground represented by P_2 is also depicted on photographs 1 and 3; it can be transferred to photographs 1 and 3 by comparison of photographic detail or stereoscopically (see Chapter 3). Similarly, P_1 and P_3 can be transferred to photograph 2. The process of transferring principal points and joining them up is known as **base lining** the aerial photographs; and the lines P_1–P_2 or P_2–P_3 on photograph 2, are known as **base lines** (they are not necessarily exactly collinear).

Now *b* represents a point of detail on the ground, which is depicted on photograph 1 to the right of the principal point P_1; that is, it is in the right-hand half of the photograph. On photograph 2, *b* still appears to the right of P_1 of course, because both *b* and P_1 represent actual places on the ground and must retain their same relative positions, but it is now on the left-hand side of the photograph; there has been an apparent movement of both P_1 and *b*. If photographs 1 and 2 are regarded as equivalent to the two different views an observer would obtain from the Air Stations 1 and 2, then the apparent movement of *b* relative to the photograph centres is its parallax. Suppose the two photographs are set up for viewing stereoscopically and the right eye is closed and the left photograph is inspected through the stereoscope, the position of *b* relative to the centre of the photograph in the field of view (i.e. P) will be b', as shown in Fig. 6. Similarly, if the left eye is closed and the right photograph inspected, the position of b relative to the centre of the photograph in the new field of view (again represented by P) will be b''. The distance b'–b'' is the parallax of point b; it is sometimes referred to as the **absolute parallax** of the point.

Fig. 6.

Parallax is measured parallel to the base line; hence if, in Fig. 5, perpendiculars are dropped from b to the base lines on both photographs 1 and 2, the parallax of b is given by the sum of the distances L_1 and L_2. If the ground is perfectly horizontal, the parallax of all points is the same as the length of the air base, i.e. the distance P_1–P_2.

DIFFERENCE IN PARALLAX

A difference in the elevation of a point affects its parallax. Consider the vertical section Fig. 7, in which L represents the camera lens, T and B respectively the top and bottom of a tall vertical chimney, P the principal

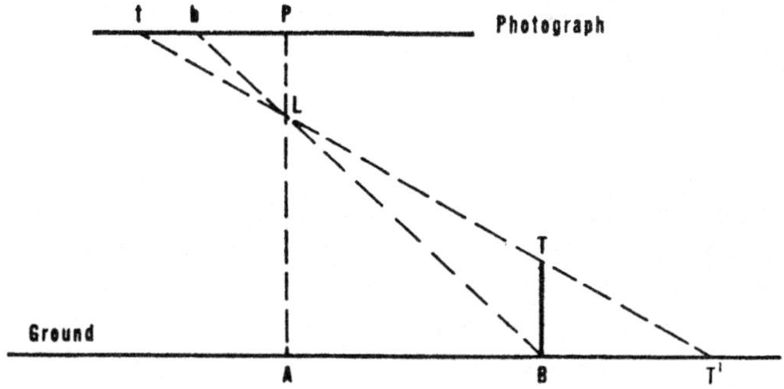

Fig. 7.

point on the photograph, and A the point on the ground corresponding to the principal point on the photograph.

If the ray of light BL is considered, it will be seen that B will be recorded at point b on the photograph. Any details on the ground, immediately adjacent to B, will be recorded on the photograph, in their correct relative position, immediately adjacent to b. If however the ray TL is considered, it will be seen that it is coincident with the ray $T'L$; thus the point T will be recorded on the photograph at point t, coincident with the ground details around T'.

Although T is vertically above B and on a map would have the same planimetric position, on the aerial photograph its image t is displaced radially from the principal point by a distance b–t. This displacement results from the elevation difference between T and B; in fact, on a truly vertical photograph, all elevation differences produce displacements radial from the principal point. The phenomenon is known as **relief displacement.**

The effect of relief displacement on parallax is illustrated in Fig. 8. The photographic images of the top and bottom of a vertical chimney are represented by t and b respectively. Notice how the line b–t, the length of which represents the relief displacement of t relative to b, is radial from P_1 on

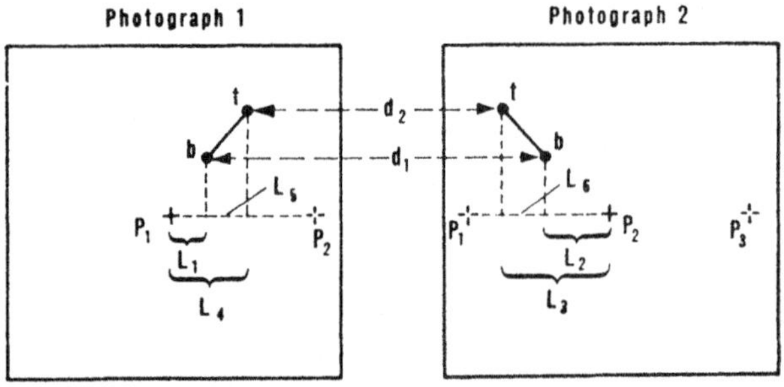

FIG. 8.

photograph 1 and P_2 on photograph 2. (The line b–t on the photograph is of course the photographic image of the chimney.)

The parallax of b is $L_1 + L_2$, and that of t is $L_3 + L_4$ (see section on "Parallax" above). As L_4 is greater than L_1 and L_3 is greater than L_2, it is obvious that the parallax of t is greater than that of b. The higher point has the greater parallax.

Thus a difference in elevation, in addition to producing radial displacement, also produces a difference in parallax. It is this difference in parallax on the photographs that produces an illusion of three dimensions when the photographs are viewed stereoscopically (see Chapter 3) and that provides a measure of the elevation difference (see Chapter 10).

In Fig. 8, it can be seen that the difference in parallax between t and b is given by the difference between $L_3 + L_4$ and $L_1 + L_2$, which is $L_5 + L_6$ Now if the photographs are correctly aligned, i.e. with their base lines collinear, it can be seen that the parallax difference could be obtained merely by measuring very accurately the distance d_1 between the two b's and the distance d_2 between the t's (see Fig. 8) and subtracting one from the other. This is in fact the basis of height finding and contouring on aerial photographs, which will be considered in Chapter 10, "Photogrammetry for Geologists".

Photogrammetry is the science of obtaining reliable measurements by means of photography. Planimetric mapping by photogrammetric methods is based upon the **radial assumption,** which is that all displacements on a photograph resulting from elevation differences or small amounts of tilt are radial from the principal point; this is true for relief displacement but not exactly true for tilt displacement. If the tilt is slight, however, and if the ground relief is only a small fraction of the flying height, the error is acceptable, and the radial assumption is justified. Under these conditions, the angle subtended at the principal point by any two points on the photograph is correct, even though they are at different elevations and have been displaced different amounts. This means that, for mapping purposes, an aerial photograph can be used in much the same way as a *round of angles* obtained with a theodolite that has been set up at the point on the ground equivalent to the principal point of the photograph.

For the discussion of stereoscopy which follows, it is important to remember that, with the photographs correctly set up, as in Fig. 8, the distance between the photographic images of particular points on two consecutive photographs is related to the elevation of those points. The higher points have less separation, and the difference in separation (difference in parallax) is proportional to the difference in elevation.

CHAPTER 3

Stereoscopy

THE act of perception is a mental process; the mind invents a model to fit the data with which it has been provided. If Fig. 9A is looked at persistently, it will appear as a book-like object alternately opened towards and away from the observer. The mind has insufficient data to decide between these two possibilities. Such illustrations are known as **alternating figures** (Vernon, 1962). As soon as extra data are added to the illustration to make one interpretation more probable than another, the observer tends to "see"

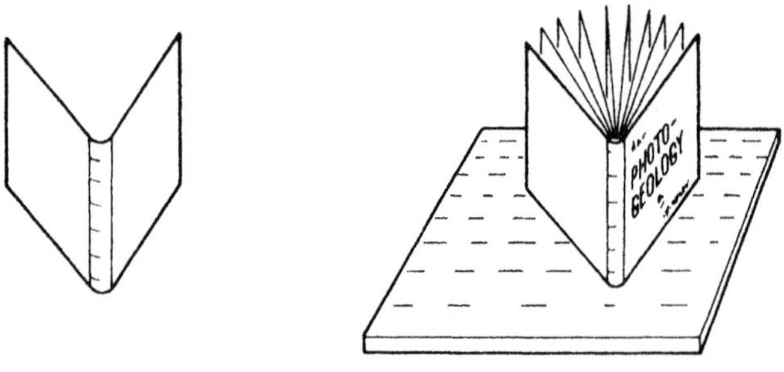

A B

FIG. 9.

it in that way. Thus in Fig. 9B, the addition of pages and a table top make the illustration appear permanently as a book opened away from the observer.

Binocular stereoscopic vision is not absolutely essential for the production of a three-dimensional mental model. The film technique "Cinerama" produces an illusion of three dimensions, of varying intensity, by the use of a curved screen and stereophonic sound. The stereophonic sound assists in the production of the mental model. The strength of the three-dimensional effect is the resultant of all the data presented to the mind.

A simple experiment may convince the reader of this. If a number of objects on a table are looked at with only one eye, after a short time, the field of view appears flat. If then the head is moved from side to side, an increase in the apparent solidity of the field of view occurs. The effect of moving the head from side to side is to provide the mind with additional data (viz. change in perspective of the objects with the movement of the head); these extra data lead to the production of a three-dimensional, rather than two-dimensional, mental model.

There are many factors, such as the relative sizes of objects, the distribution of shadows, and perspective, which affect our perception of three-dimensional space. If all the various factors tend to produce the same three-dimensional effect in the mind, then the effect is at its maximum strength. If, however, the different factors tend to give a different indication of what the three-dimensional mental model should be, then the effect of the factors will be subtracted from one another; either no solid model will be "seen", or, if one set of factors produces a stronger indication than the other, a less convincing solid model will be "seen". This will be discussed more fully under "Pseudoscopic Vision".

STEREOSCOPIC VISION

For most people, the factor that produces the strongest three-dimensional effect is stereoscopic vision. Stereoscopic vision depends upon having two view-points (the eyes), set about 2·5 in. (6·3 cm) apart in the head; this distance is known as the **eye-base** (see Fig. 10).

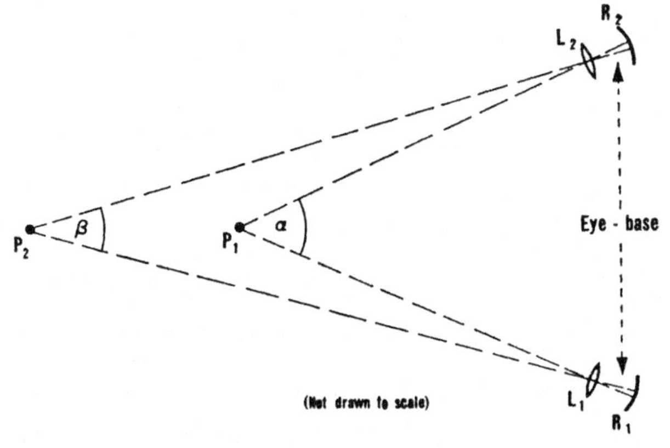

FIG. 10.

In Fig. 10, P_1 and P_2 represent two pencils held vertically in front of the eyes; P_1 is about 10 in. (25·4 cm) from the eyes and P_2 a little further; L_1 and L_2 represent the lenses and R_1 and R_2 represent the retinas of the left and right eyes respectively.

Consider what occurs in the process of looking at P_1. Firstly, the eyes converge so that each is aimed directly at the pencil; secondly, each eye focuses on the pencil; and thirdly, the two distinct and slightly dissimilar images received (one for each eye) are used for the production of a single, three-dimensional, mental model. The first two stages are known as **convergence** and **accommodation** respectively.

The angle of convergence is defined as the angle subtended by the eye-base at the object observed; it is angle α for P_1 and β for P_2 in Fig. 10.

The convergence of the eyes looking at the pencil indicate to the mind how far away the pencil is, and at what distance the eyes should be focused. In practice, as the gaze moves from object to object, focus and convergence are altered simultaneously so that they become virtually interrelated.

When the gaze is concentrated on P_1, the eyes converge and focus on that point so that the two images of P_1 (one for each eye) are brought to a focus on corresponding points on the retina of each eye. A single image of P_1 is seen; this is known as **fusion** of the images, or stereoscopic fusion.

As the required angle of convergence for P_2 is less than for P_1, it follows that, when the gaze is concentrated on P_1, the light rays from P_2 fall on non-corresponding points of the retina of each eye. The mind is able to fuse different pairs of images simultaneously, if the amount of non-correspondence of the retinal images is not too great; if it is too great, however, simultaneous fusion of the image pairs is impossible, and double images are seen.

The amount of non-correspondence of the retinal images is interpreted by the mind as an indication of the distance that P_2 lies behind P_1; this is so even though both P_1 and P_2 are in fusion. Thus, a three-dimensional mental model is built up.

If the gaze is directed to P_2, the images of P_2 fall on corresponding points of the retinas but those of P_1 produce non-corresponding retinal images, in the opposite sense to that above; P_1 is thus interpreted as being in front of P_2.

It is possible to simulate the binocular observation of the two pencils by means of a diagram. If in Fig. 10, pencil 1 is grey and pencil 2 is black, then the left and right sides of Fig. 11 may be regarded as respectively equivalent to the views of the left and right eyes in Fig. 10. If this is not easy to understand, imagine the rays of light from P_1 and P_2 passing through a piece of transparent paper held perpendicular to the plane of the figure (Fig. 10); the positions of the two grey and the two black shapes in Fig. 11 are respectively equivalent to the places where the rays from P_1 and P_2 to the eyes pass through the paper. The two grey shapes in Fig. 11 are closer together than the two black, as will be seen to be necessary from a consideration of Fig. 10. No significance should be attached to the fact that the black shapes are to the right of the grey on both sides of Fig. 11, instead of the grey lines being in the centre, as would be expected from an inspection of Fig. 10. The figure has been drawn thus merely to facilitate fusion; it is the amount by which the separation of the grey shapes differs from that of the black (viz. the stereoscopic parallax), which is significant.

The separation of the grey shapes in Fig. 11 is less than that of the black; if, therefore, the left and right eyes are respectively concentrated on the left and right grey shapes, and the images are fused, there will be a non-correspondence of retinal images for the black shapes, which will lead to an interpretation that there is a single black shape behind a single grey one.

The stereo-diagrams in Figs. 11, 12, and 13 have been designed to be viewed with a pocket stereoscope. Readers experienced in the stereoscopic use of aerial photographs will probably be able to fuse them without the assistance of a stereoscope. A method of developing this ability is described in the section below entitled "Viewing Stereopairs without a Stereoscope".

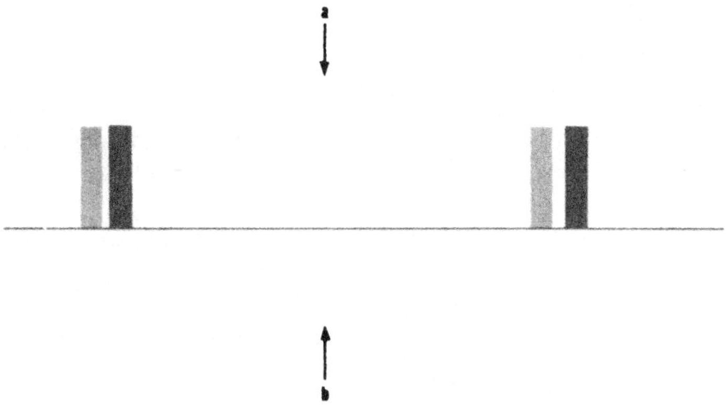

FIG. 11.

It may be necessary to persist in the stereoscopic study of Fig. 11 for a few minutes before the three-dimensional effect becomes apparent; people who have one eye stronger than the other or scientists who have spent much time on microscopic work in which they have used only one eye, may have to persevere longer than others before they can fuse them satisfactorily.

In Fig. 11, there are no shadow or perspective effects to assist in the production of the stereomodel, and thus the model appears less convincing than with a stereopair of aerial photographs. This illustration does demonstrate however that what is "seen" is not necessarily what exists in the field of view of the observer; there are four shapes on a plain piece of paper, but two shapes are "seen", one behind the other, in apparently three-dimensional space.

If the shapes representing the more distant (black) pencil are drawn smaller than those representing the grey, perspective assists the parallax effect in the production of the three-dimensional mental model. Thus the three-dimensional illusion received from Fig. 12A should be stronger than that from Fig. 11. The effect depends on the observer assuming that the two objects in his stereomodel are really the same size; the apparently smaller one is then interpreted as being further away.

If the black shapes are larger than the grey, as in Fig. 12B, the perspective effect operates against the parallax effect and the three-dimensional effect is

FIG. 12.

reduced. If, therefore, Figs. 12A and B are studied under a pocket stereoscope, the three-dimensional illusion should be stronger for the former than for the latter.

It is now necessary to relate this discussion to the properties of aerial photographs. It will be remembered that in Chapter 2 it was shown that a difference in elevation resulted in a displacement radially from the principal point of a photograph. Thus, if a vertical chimney is depicted on two consecutive photographs, the photographic image of its top *t* is displaced relative to its bottom *b*, radially from the principal point of each photograph, as shown in Fig. 8. When the photographs are correctly set up for stereoscopic viewing, the two images of the top are nearer together than the two images of the bottom (difference of parallax), just as in Fig. 11 the two grey shapes

FIG. 13.

are nearer together than the two black. The result is as illustrated in Fig. 13 in which the pairs of dots *t*, *m* and *b* represent the two images of the top, middle and bottom points of the chimney respectively. When the dots in Fig. 13 are fused, *t* appears to be above *m*, and *m* appears above *b*. What is seen therefore, when two consecutive photographs depicting the chimney are fused, is a three-dimensional stereomodel of the chimney sticking out of the ground.

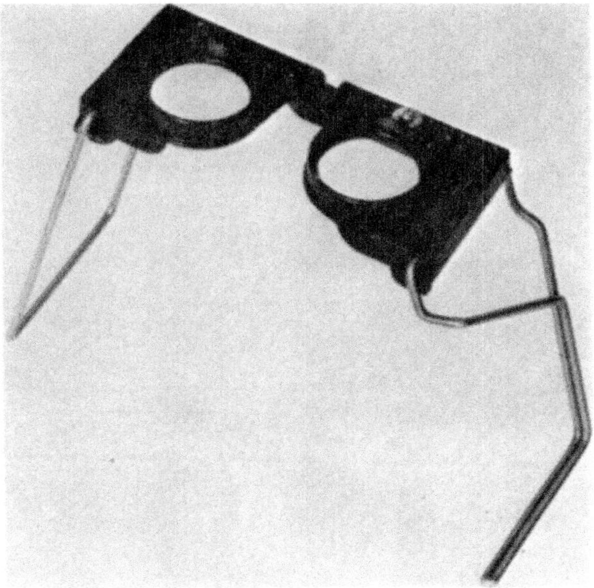

Plate Ia. Pocket stereoscope with cut-away leg.

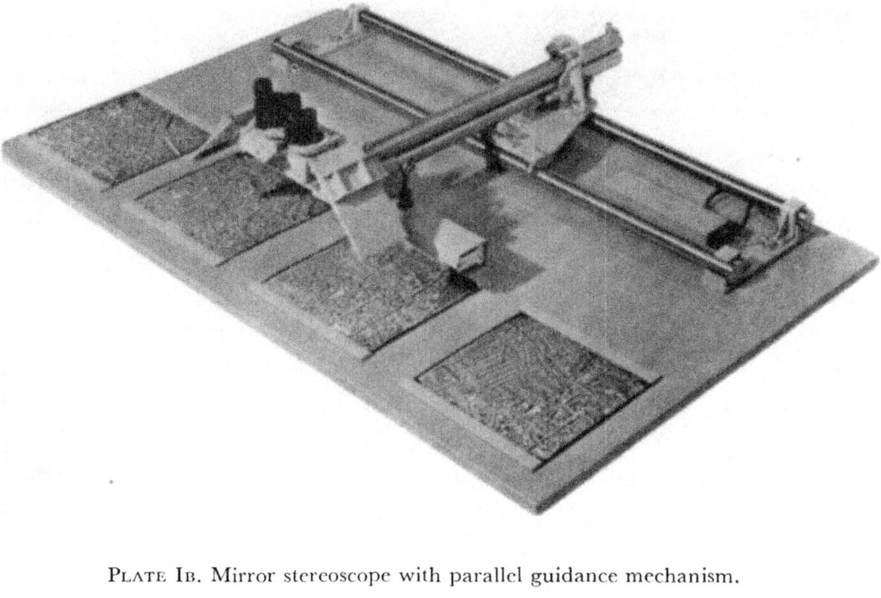

Plate Ib. Mirror stereoscope with parallel guidance mechanism.

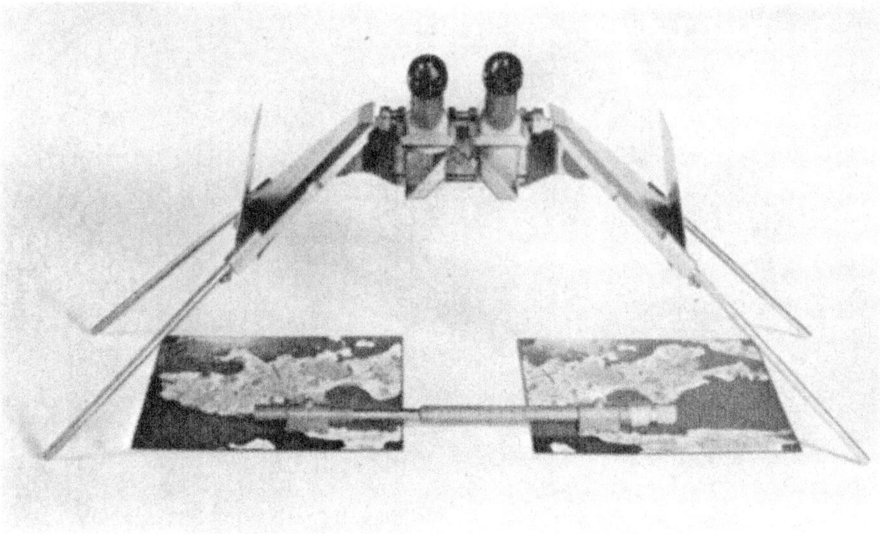

Plate IIa. Mirror stereoscope and stereometer.

Plate IIb. Ground-level photograph of permeation gneiss, Nigeria.

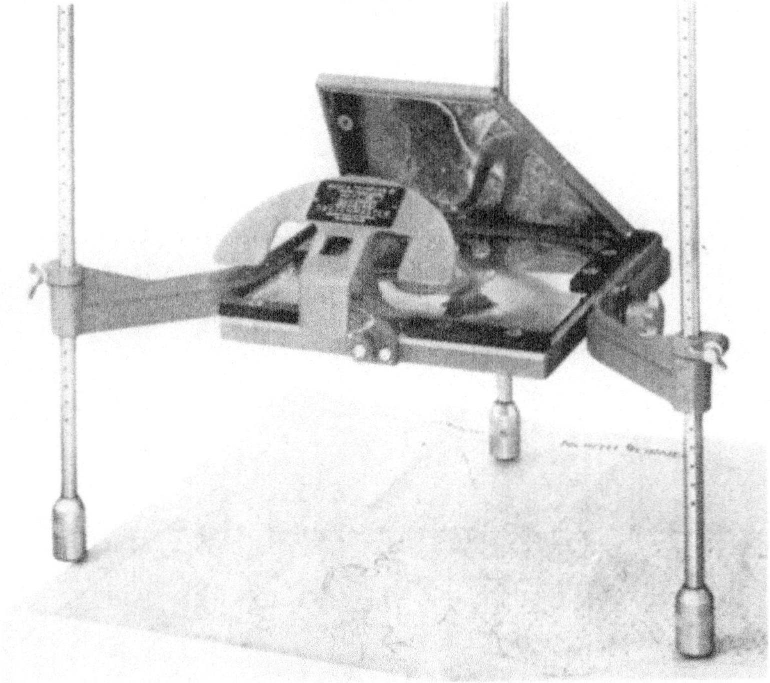

PLATE IIIA. Sketchmaster.

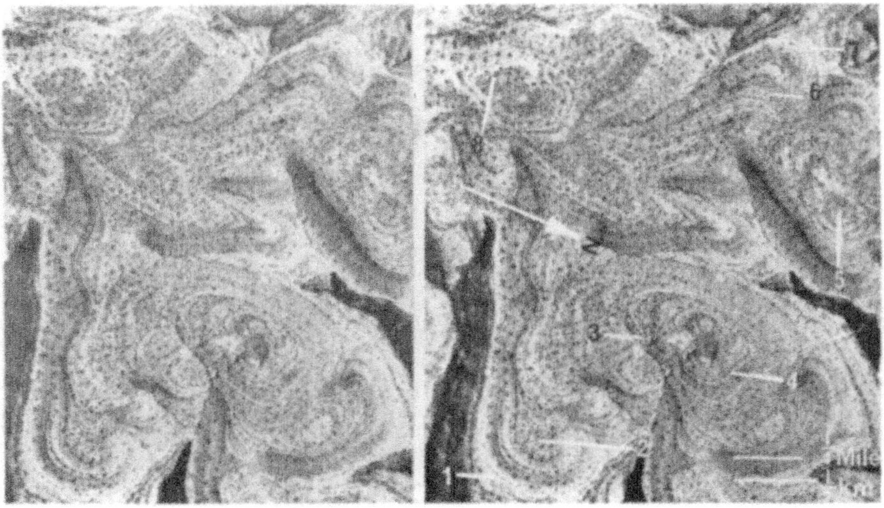

PLATE IIIB. Stereopair. Horizontally bedded limestone and shale, Somali Republic.

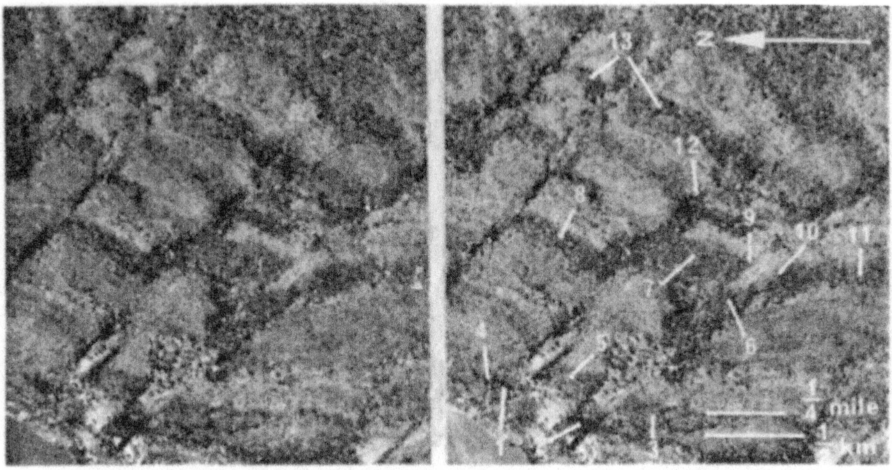

Plate IVa. Stereopair. Faulted, dipping, Karroo sandstones, grits, and marls, U.R. Tanzania.

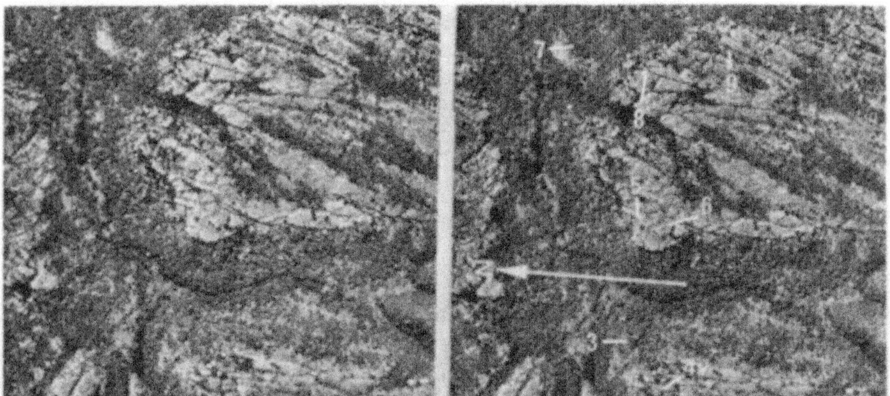

Plate IVb. Stereopair. Horizontally bedded, much-jointed sandstone, U.R. Tanzania.

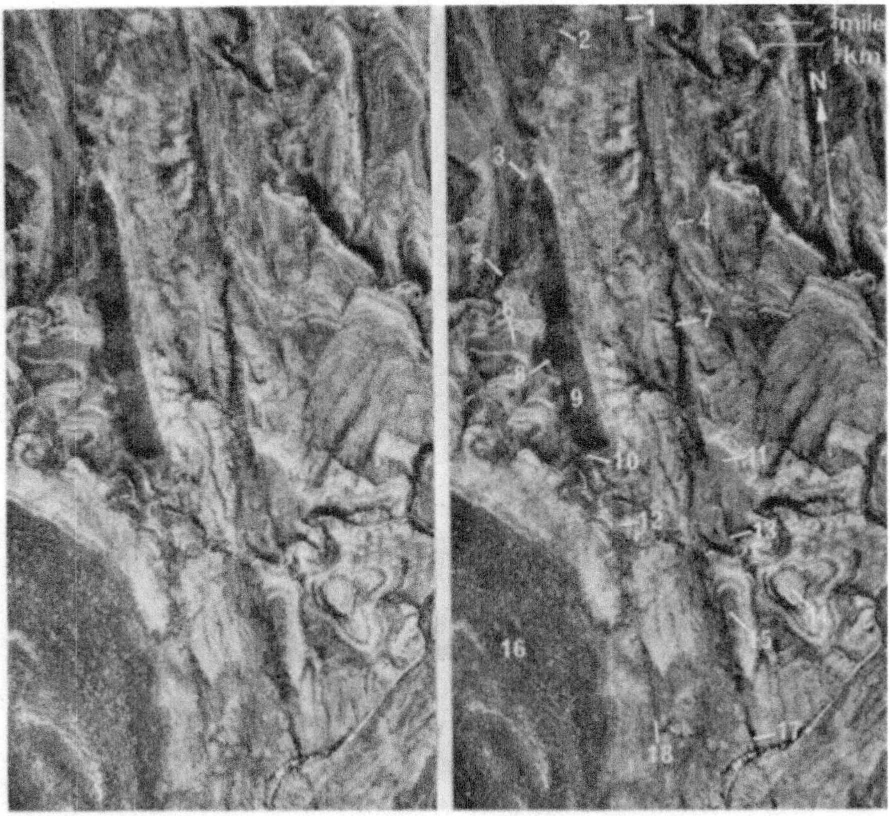

Plate V. Stereopair. Faulted Karroo mudstones, sandy mudstones, and grits, Zambia.

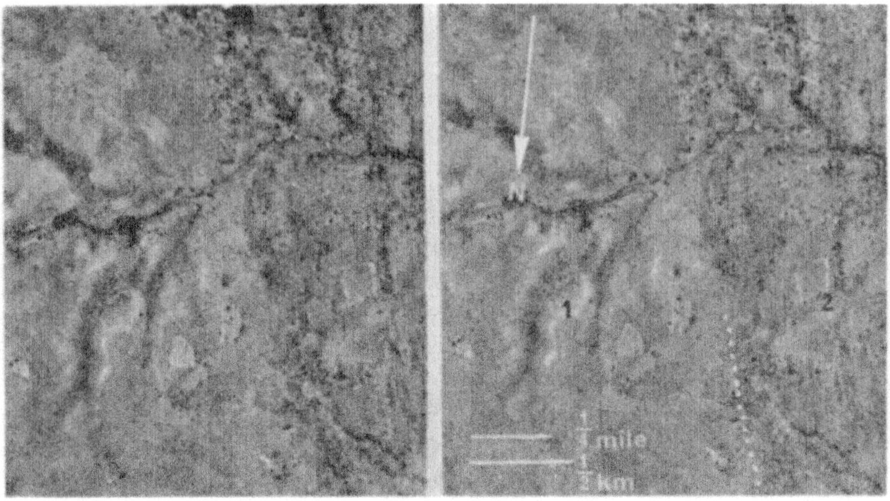

PLATE VIA. Stereopair. Soil-covered boundary between granite 1 and metasediments 2, Nigeria.

PLATE VIB. Stereopair. Boundary between granite and metasediments (limestones, light-toned; quartzites, dark-toned), Somali Republic.

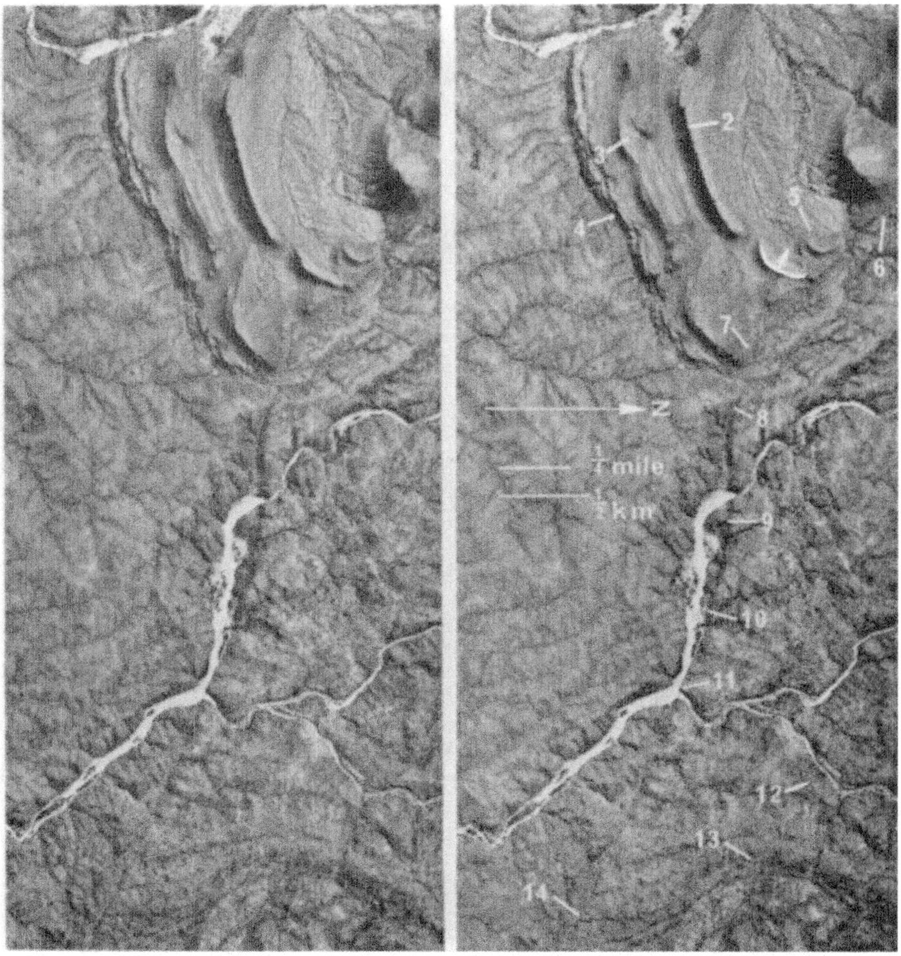

PLATE VII. Stereopair. Folded limestones and shales, and Basement Complex, Somali Republic.

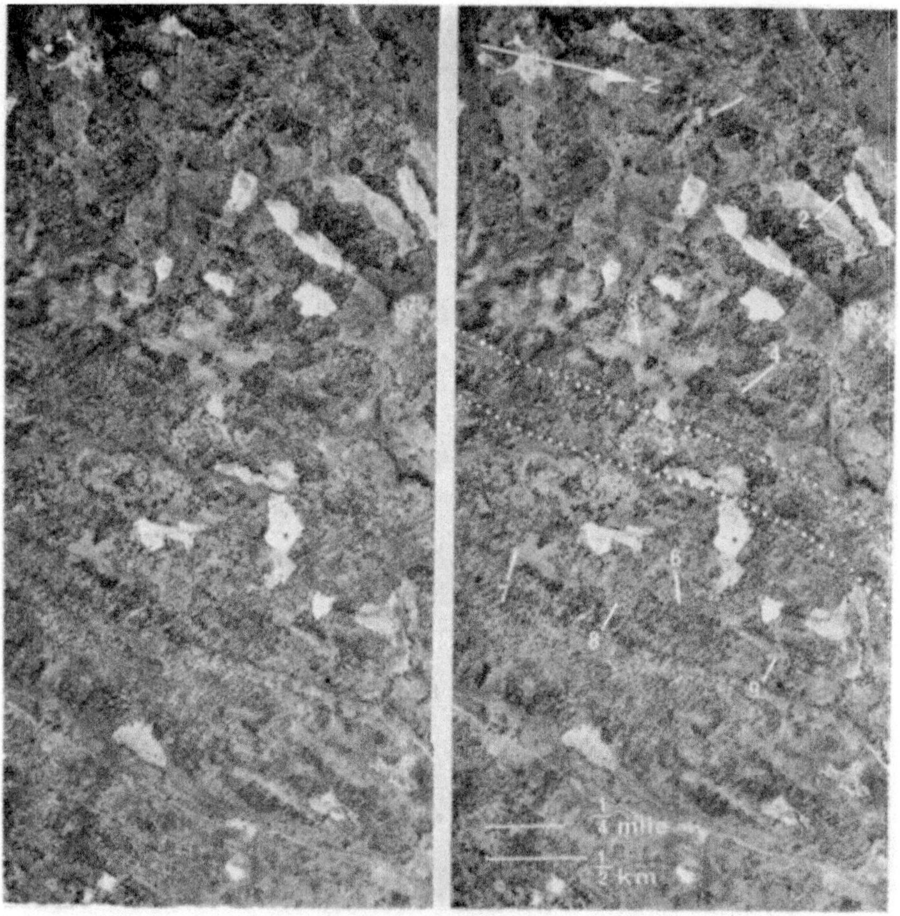

PLATE VIII. Stereopair. Pebbly, semi-pelitic schists and phyllites, Nigeria.

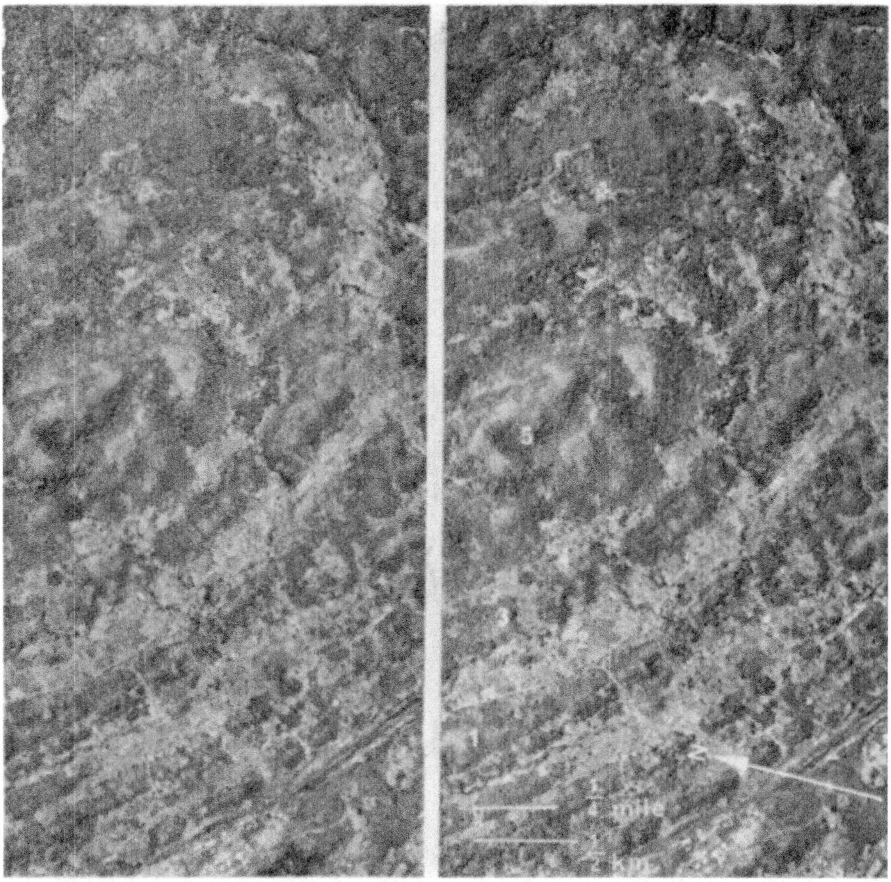

PLATE IX. Stereopair. Folded dolomitic limestones 2, 4 and hornblende gneisses 1, 3, 5, U.R. Tanzania.

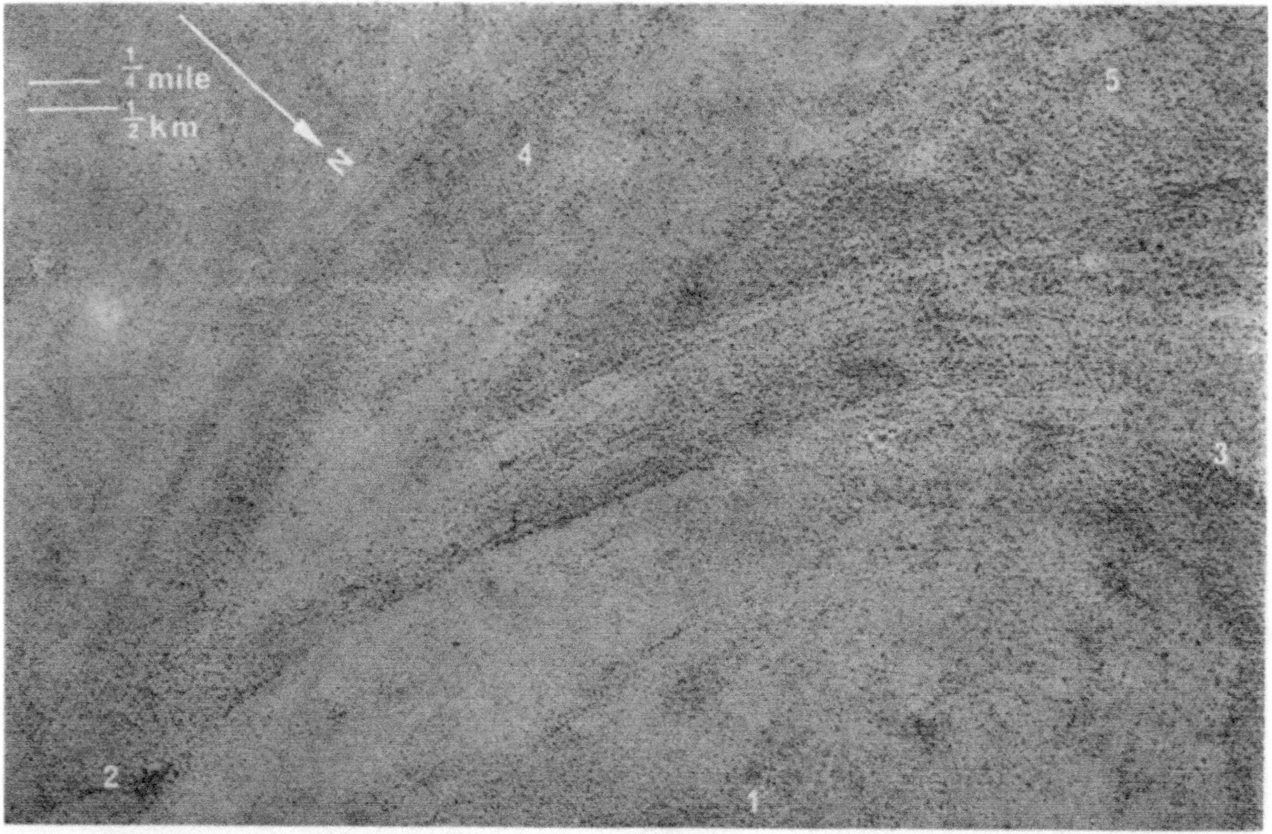

PLATE X. Soil-covered, folded (?) metasediments, Bechuanaland.

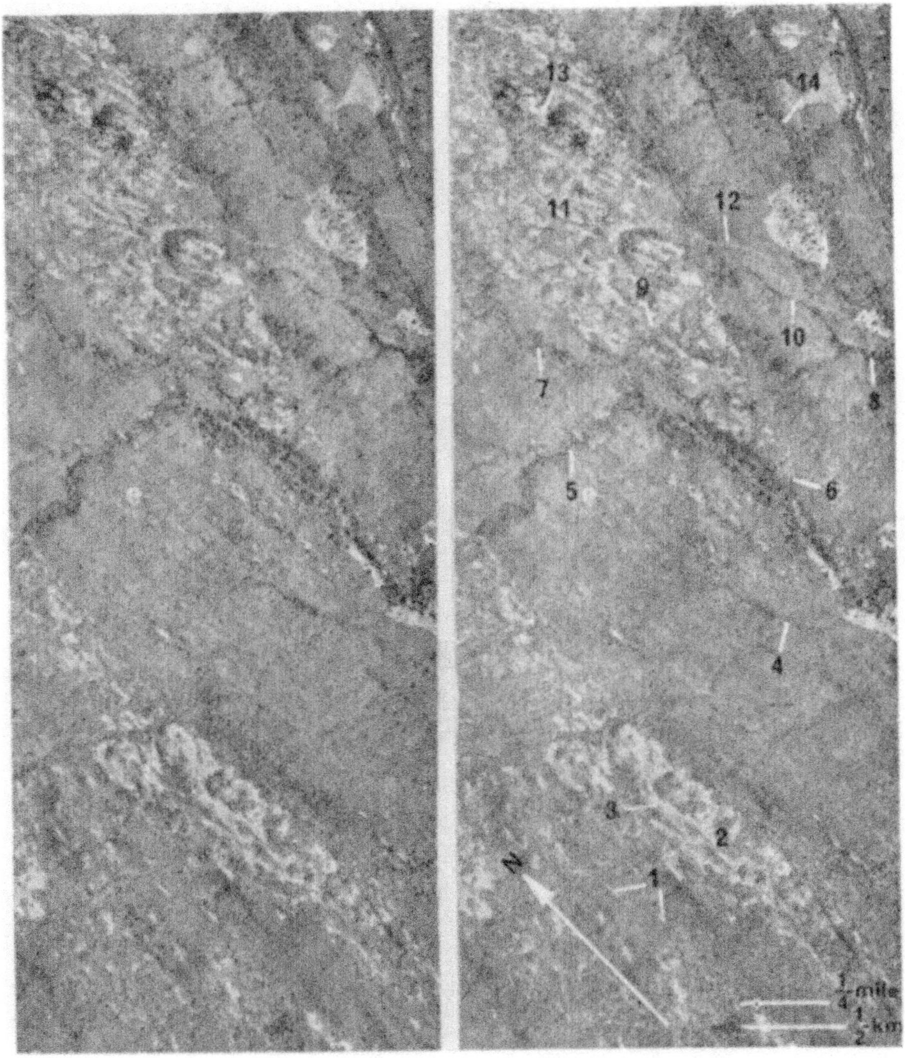

PLATE XI. Stereopair. Granite 11, permeation gneiss 2, and metasediments, Nigeria.

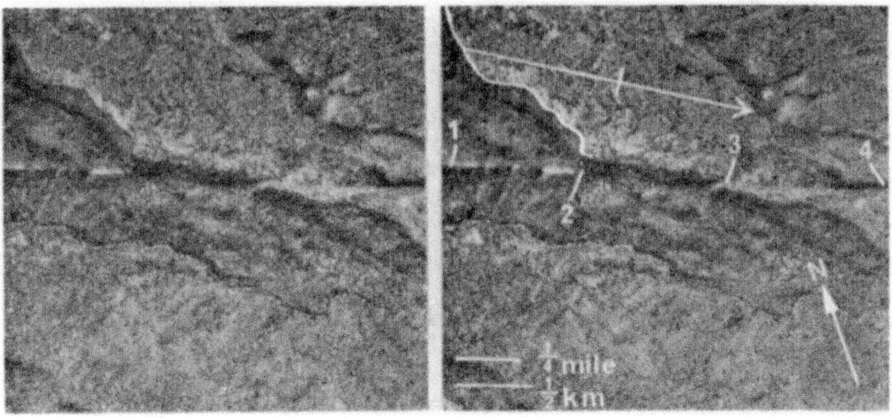

PLATE XIIA. Stereopair. Dolerite dyke cutting Upper Karroo grits and mudstones, Zambia.

PLATE XIIB. Stereopair. Granite 3, British Guiana.

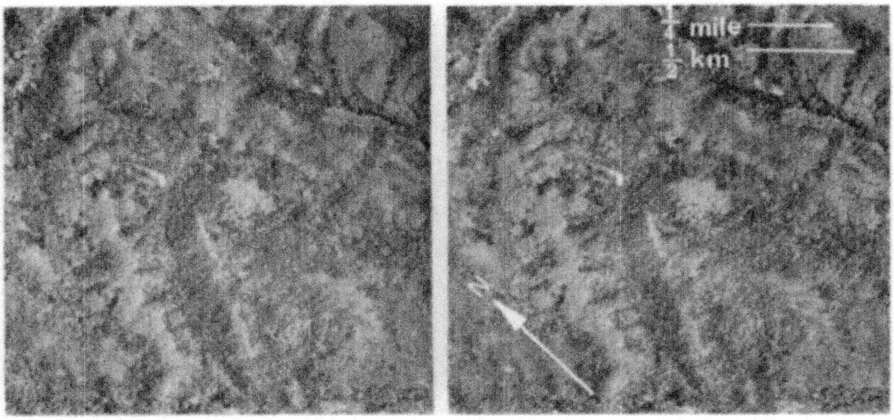

PLATE XIIIA. Stereopair. Folded semi-pelitic schists and phyllites, Nigeria.

PLATE XIIIB. Stereopair. Plunging folds in quartz-schists, and granite, Nigeria.

PLATE XIVA. Stereopair. Folded and faulted hornblende gneisses, U.R. Tanzania.

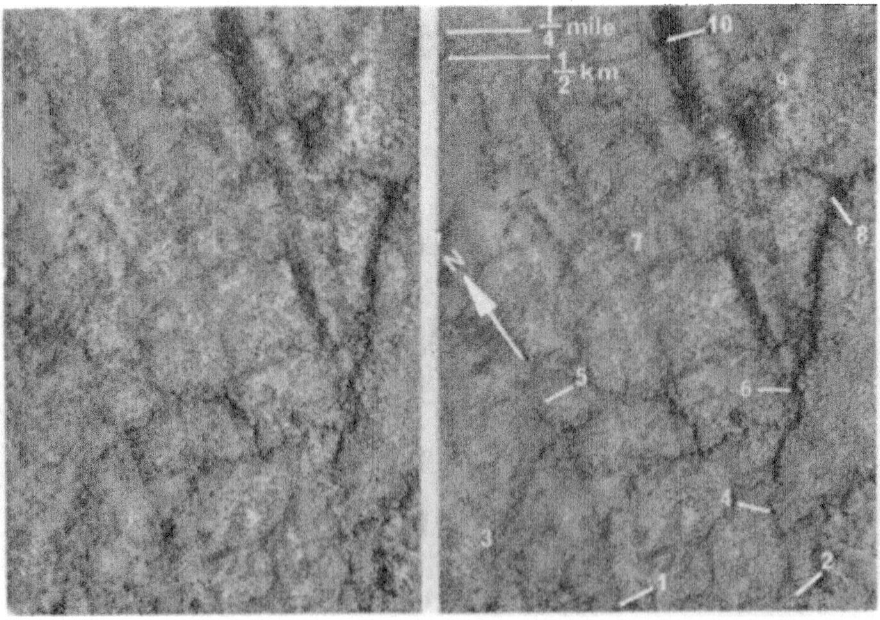

PLATE XIVB. Stereopair. Faulting in metasediments, granulite 9, pelitic schist and phyllite 7, and granite 3, Nigeria.

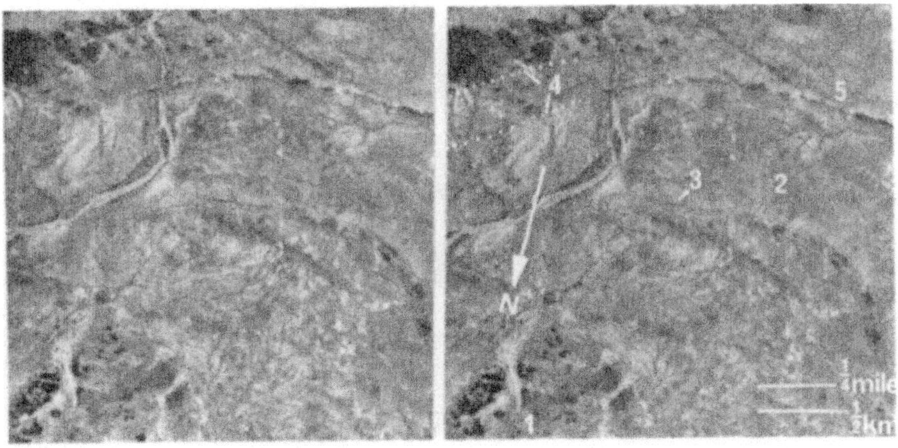

PLATE XVA. Stereopair. Peridotite 1, gabbro 2, and monzonite 5, Somali Republic.

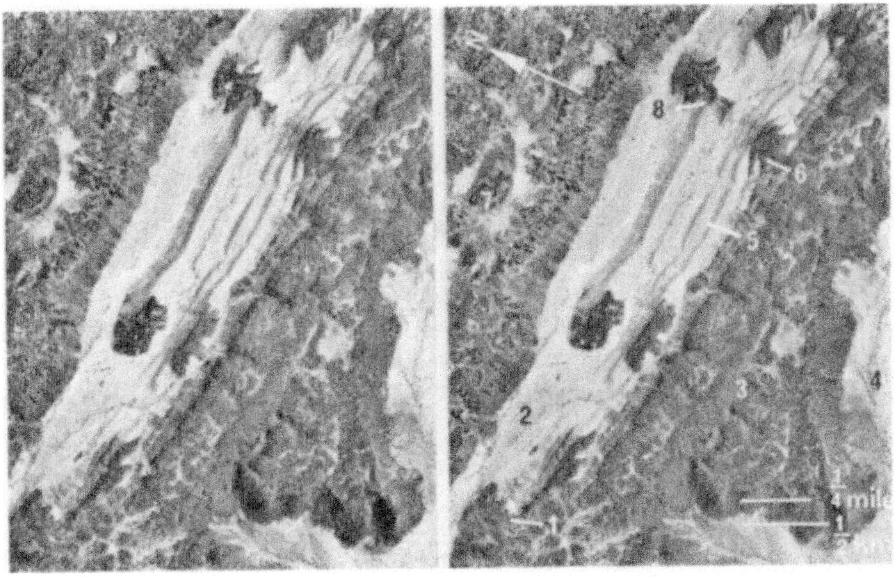

PLATE XVB. Stereopair. Basaltic lava 3, 7 and limestones 5, Somali Republic.

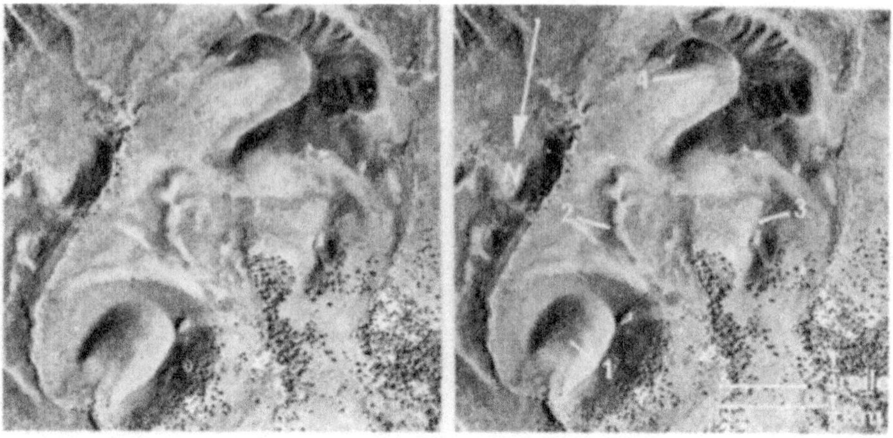

Plate XVIa. Stereopair. Partly eroded volcanoes, Kenya.

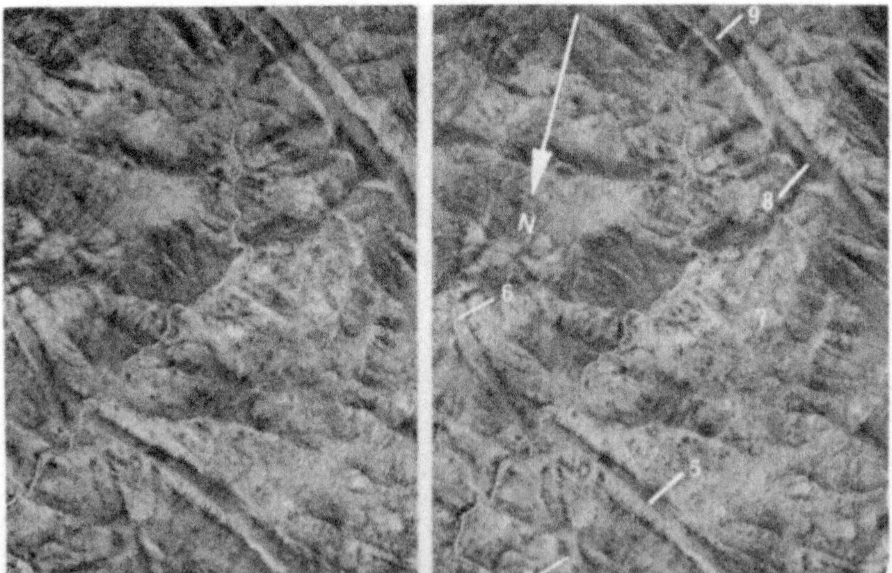

Plate XVIb. Stereopair. Basic dykes, Somali Republic.

The arguments that we have applied to different points on the chimney are relevant to all points on the common overlap of two consecutive photographs. The higher a point is on the ground, the greater is its relief displacement, and the closer together are the corresponding images on the two photographs. The closer together the corresponding images are, the higher the fused image appears on the stereomodel. Thus the stereomodel reproduces the relative differences in elevation that exist in reality.

VIEWING STEREOPAIRS WITHOUT A STEREOSCOPE

It is possible to fuse two images stereoscopically without the use of a stereoscope. Hold a stiff piece of cardboard, about 10 in. (25·4 cm) square, perpendicular to the paper along the line *a–b* in Fig. 11, and put your nose on the upper edge of the cardboard, so that your left and right eyes can look only at the corresponding sides of the figure. (It is important that the incident light should come from in front of you so that both sides of the figure are equally illuminated.) After staring for a few moments, the left and right images should fuse.

The two sides of the diagram in Fig. 11 are fairly widely separated; thus when they are fused, the lines of sight of the eyes are nearly parallel, and the angle of convergence is small. For people practised in stereoscopy, fusion can still be obtained if the separation of the two sides is equal to, or even slightly greater than, the length of the eye-base. In such a case, the lines of sight are parallel or diverging, and the angle of convergence is zero or negative. This confirms that the three-dimensional effect is dependent not upon the angle of convergence but upon the difference in parallax between the pairs of images.

Most people will have no difficulty in fusing the objects (blue and black lines) in Fig. 11 without a stereoscope, but they will have difficulty in focusing them, when fused. This is because, when the lines of sight of the eyes are nearly parallel, the small angle of convergence suggests to the mind that the objects viewed are far off; the eyes therefore focus at infinity, and the object appears out of focus. In order to focus as well as fuse the objects in the figure, it is necessary to learn to dissociate the processes of focusing and converging the eyes. This can usually be achieved with practice, but it is not really worth the effort. A pocket stereoscope enables a stereopair or stereodiagram to be studied closely, at a distance of about 5 in. (12·7 cm), whilst the eyes are focused at infinity.

PSEUDOSCOPIC VISION

If the positions of the two photographs depicted in Fig. 8 were exchanged, but their orientations kept unchanged, the distance between the two images *t* would be greater than that between the two images *b*. Under these conditions, the parallax difference is reversed, and if the photographs were examined stereoscopically, the observer would experience pseudoscopic vision, in which all the hills were represented by basins, and all the rivers by ridges.

Stereopairs are not, in all circumstances, essential for the production of a pseudoscopic effect; certain single photographs produce the illusion of inverted relief when orientated so that the direction of the shadows is away from the observer. If such a photograph is rotated, the relief suddenly appears normal; if it is then rotated slowly back to its original position, the relief probably continues to appear normal.

This effect is apparently related to an assumption, made by the observer, as to the direction of the light at the time the photograph was taken. It is as though the observer assumes that the light came from the direction he is facing. If this assumption is wrong, inverted relief may be seen. With the assumed direction of light remaining constant whilst the photograph is rotated, the appearance of normal relief suddenly results. Once the normal relief has been "seen", the true direction of the incident light at the time of photography is established, and further rotation of the photograph may not destroy the appearance of normal relief.

When viewing a stereopair, the effect of parallax difference outweighs all others in the production of the three-dimensional illusion; most people do not experience pseudoscopic vision when viewing a correctly set-up stereopair, whatever the direction of shadows. Nevertheless, the stereomodel appears at its best, and the most convincing relief is "seen", when the shadows on the photographs are aligned towards the observer.

VERTICAL EXAGGERATION

If the length of the eye-base in Fig. 10 is increased, the angles α and β, and the difference between these angles, are all increased. The difference between the angles α and β is known as the angle of differential parallax, and it is a measure of the non-correspondence of the retinal images of P_2, when the gaze is concentrated on P_1. Thus an increase in the eye-base results in an increase in depth perception, i.e. in an increase in the apparent distance of P_2 behind P_1.

If the eye-base is kept constant and the distance between the eyes and P_1 increased, the angles α and β and the angle of differential parallax all become smaller; there is then a decrease in depth perception, and the apparent distance of P_2 behind P_1 becomes smaller.

In the stereo-diagrams, Figs. 11 and 12A, the equivalent to the angle of differential parallax is the difference in the distances separating the two blue and two black shapes. If the difference of separation is increased, the apparent distance of one shape behind the other in the fused stereomodel is increased, and vice versa.

The minimum variation in the angle of differential parallax that can be detected by the eyes is of the order of 30 seconds of arc. An object about 1400 ft (427 m) away from a man with an eye-base of 2·5 in. (6·34 cm) subtends this angle of 30 seconds of arc, and therefore judgements of distances greater than this cannot be stereoscopic at all; it can be done just as well by a man with one eye. Thus a man flying several thousand feet above the ground cannot use his stereoscopic vision in determining relief. Any impression of relief that he obtains must be the result of other causes such as light and shadow effects (Trorey, 1950).

In order to keep the depth perception constant whilst increasing the viewing distance, it is necessary that the eye-base should be increased in proportion to the viewing distance, i.e. that the ratio of eye-base to viewing distance should be constant.

Each photograph of a stereopair may be regarded, for the purpose of this discussion, as representing what the eye would see at the position at which the photograph was taken (the air station). Thus, when the observer studies a stereopair, his effective eye-base is not his natural eye-base, which is of course fixed, but the air base of the photography. If the ratio of the air base to the height of the aeroplane above the ground at the time the photographs were taken is the same as the ratio of his natural eye-base to the viewing distance discussed above, the observer's depth perception will be unaffected. The increase in the effective eye-base will have compensated for the increase in viewing distance.

In practice, for the sake of economy, the maximum air base is usually chosen that will provide a 60 per cent overlap of adjacent photographs (see Chapter 1), without regard to the effect on the depth perception of the observer. Thus many stereomodels have vertical scales which are greater than their horizontal ones; their relief is exaggerated; the effect is referred to as **vertical exaggeration** (Miller, 1961; Verstappen, 1977).

Vertical exaggeration results in all slopes appearing to dip more steeply, and in ridges and other land forms appearing to be more clearly defined on the stereomodel, than they do in reality. Because of the exaggeration of relief, low, rounded, and subdued topographic features, which would normally be overlooked in the field and which may be indicative of geological structures, may be observed and plotted. Vertical exaggeration is therefore an advantage in photogeology.

On occasions, stereopairs will be found having a very large overlap (very short air base). These have usually resulted from some error in setting the camera, which has caused it to take two photographs in rapid succession.

Stereomodels formed from such stereopairs show little or no vertical exaggeration and may even exhibit vertical diminution (negative exaggeration). If the air base were of zero length, the photographs would be identical, there would be no parallax difference between them and no relief would be visible on the stereomodel.

STEREOSCOPES

Most geologists use a mirror stereoscope in the office and a pocket stereoscope in the field.

A pocket stereoscope (Plate IA) consists of two simple lenses mounted on a metal frame; the frame has short folding legs, so that, when not in use, the stereoscope can be kept in a pocket. The length of the frame can be adjusted to suit the eye-base of the user.

The best separation for corresponding points of detail on a stereopair, when viewing with a pocket stereoscope, is slightly less than the eye-base of the observer, about 2·25 in. (5·72 cm). If two 9 in. (22·9 cm)-square aerial photographs, having 60 per cent overlap, are put side by side with their edges touching, in their correct relative orientation, the distance between points of corresponding detail is 5·4 in. (13·7 cm) plus the width of any edge or titling strip. To view such photographs with a pocket stereoscope therefore, it is necessary to overlap them until the corresponding photographic images are at a suitable separation. When this is done, a strip about 2·25 in. (5·72cm) wide may be viewed stereoscopically; the rest is obscured by the overlapping photograph. To view the whole overlap, it is necessary to bend back the edge of the overlapping photograph.

The reader is recommended to persevere with his efforts at using a pocket stereoscope, because once the knack has been learned, he will be able to use the instrument in the field at a moment's notice. Many geologists, who do not develop this aptitude, end by not using aerial photographs stereoscopically in the field at all. Their field mapping suffers, and they lose the advantage of frequent correlations between the stereomodel and the ground view, which would have assisted them in future photogeological interpretations.

The instrument illustrated in Plate IA has a cut-away leg. This modification provides more space under the instrument for writing; it is such a convenience for annotation purposes that it is difficult to understand why all instruments are not manufactured in this form.

Different models of pocket stereoscopes have different magnifications. The one illustrated, which is fairly typical, has a lens magnification of $\times 2$.

Mirror stereoscopes are illustrated in Plates IB and IIA; their optical system is as shown in Fig. 14. Light rays from the photographs are reflected first by large, surface-silvered mirrors set at 45° to the horizontal and then again by small, surface-silvered mirrors set parallel to the large ones. After reflection by the small mirrors, the rays are parallel to their original direction (i.e. vertical), but are separated by a distance determined by the separation of the small mirrors. The separation of the small mirrors is either adjustable or preset to approximately the eye-base of the average observer.

It can be seen that, with this system, the separation of the photographs, when they are set up for stereoscopic viewing, is determined by the separation of the large mirrors.

The optical distance between the eyes and the photographs consists of the sum total of the distances between the eyes and the small mirrors, the small mirrors and the large mirrors, and the large mirrors and the photographs; it is thus very great compared with that for the pocket stereoscope. A stereo-pair viewed from such a distance would appear very small. To overcome this, lenses are inserted between the eyes and the small mirrors. If higher magni-

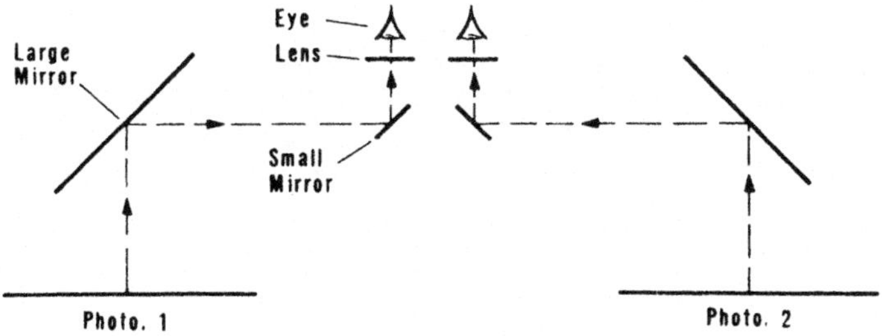

FIG. 14.

fication is required, binoculars may be inserted between the lenses and the eyes. Thus the user of a mirror stereoscope with binocular attachment has the choice of at least two different magnifications, and he can change from one to the other at will. The binoculars of most stereoscopes are angled (Plate IIA) in order to make the viewing position more comfortable.

When using high-power magnification with a mirror stereoscope, the field of view is small, and movement of the stereoscope is necessary to observe the whole overlap. If the stereoscope is moved by sliding it on its legs, small, distracting jerks occur, which result in temporary loss of the stereomodel. Therefore some form of scanning device, such as that illustrated in Plate IB, is desirable.

Because of the large number of surfaces at which the light is reflected or refracted in a mirror stereoscope, care must be taken to ensure that the instrument is kept scrupulously clean, otherwise the clarity of image will suffer.

Although some geologists like to do all their work with a pocket stereoscope, others prefer to use mirror stereoscopes on all occasions. The best arrangement is to use a pocket stereoscope in the field and in the field-camp, and a mirror stereoscope with scanning device in the office.

Whenever large areas are to be precisely annotated, a mirror stereoscope with binocular attachment is almost essential.

TRANSFERENCE OF PRINCIPAL POINTS

For some purposes (see particularly Chapter 10) it is desirable to transfer the principal point of each photograph of a strip to its two adjacent ones. When the process has been completed, each photograph, except those at the ends of the strip, will be marked with the positions of three principal points, as in Fig. 5.

The simplest way to transfer principal points is stereoscopically; a mirror stereoscope with a high-power binocular attachment should be used. The procedure is as follows:

(1) Inspect the stereopair and check that the principal point of each photograph is marked with a cross. On each photograph, note the approximate positions of the two conjugate principal points (i.e. the principal points of the two conjugate photographs), and draw circles about 0·5 in. (1·3 cm) diameter around the positions with a grease pencil. Holding one photograph in each hand, overlap them so that points of corresponding detail approximately coincide and the principal points fall within the grease pencil circles.

(2) Without altering their relative orientation, place the left and right photographs under the corresponding large mirrors of the stereoscope, and focus the eyepieces of the binoculars.

(3) Put the index finger of the left hand on a point of detail near the principal point of the left photograph and bring it to the centre of the field of view.

(4) Put the index finger of the right hand on the corresponding point of detail on the right photograph and bring it to the centre of the field of view.

(5) Remove the fingers from the photographs. It should now be possible to fuse the photographs in the immediate vicinity of the principal point.

If fusion is not possible and two separate images are being seen, choose, in the vicinity of the principal point, some clearly defined linear detail, such as a path or row of trees, and move the right-hand photograph until this linear detail is parallel to, and close to, the corresponding detail of the left-hand photograph. Then, with a sudden movement of the right-hand photograph, bring the images into fusion.

(6) The principal point of the left-hand photograph will now be seen superimposed on the stereomodel.
(7) Keeping attention confined to that part of the stereomodel very close to, but not quite at, the principal point image, bring a needle over the surface of the right-hand photograph and prick through the centre of the image of the principal point of the left-hand photograph. The main error that is likely to occur is that the right eye will look at the needle instead of the photograph just before the principal point image is pricked through; the model will then go out of fusion, and the eyes will fuse the principal point image with the needle; the needle will thus appear to go into the correct position, when it does not do so.
(8) Restudy the stereomodel to confirm that the prick-mark is in fact coincident with the principal point image. A further check is to shut the left and right eyes alternately, and to compare the minute topographic details around the principal point with those around the prick-mark.
(9) If the transference of the principal point is accurate, fill in the prick-mark with coloured-grease pencil and draw a small circle around it.
(10) It is now required to transfer the principal point of the right photograph to the left one. Repeat (1) above, and turn the resulting stereopair through 180°, so that the left photograph becomes the right, and vice versa.
(11) Repeat stages (2)–(9).

SETTING UP THE AERIAL PHOTOGRAPHS

A beginner in photogeological interpretation is advised to base line the photographs, as described in the previous section; and then to set up the photographs for stereoscopic viewing, with their base lines collinear. To do this he should:

(1) Draw a straight line in the centre of the field of view of the stereoscope and parallel to the stereoscope's axis.
(2) Place the first photograph under the left-hand large mirror with the base line collinear with the straight line (1).
(3) Place the second photograph under the right-hand large mirror with the base line collinear with the straight line (1).
(4) Put the index finger of the left hand on a point of detail on the left photograph and the index finger of the right hand on the corresponding point of detail on the right photograph and move the photographs until the fingers are in the centres of the fields of view.

(5) Check that the base lines are still collinear with the straight line (1).

(6) Remove the fingers from the photographs. The stereopair should now fuse comfortably.

After some practice, positioning the photographs will become automatic. They will merely be placed under the stereoscope in their approximately correct positions and some item of linear detail observed; then both will be moved to bring the linear detail into comfortable fusion and to align the base line with the axis of the stereoscope.

With good quality photography of fairly flat terrain, it may be possible to study a whole overlap, under high power, without further adjustment of the photograph positions. Under other conditions, a slight feeling of eye discomfort may be noticed as different parts of the stereomodel are studied. As soon as this discomfort is felt, the right-hand photograph position should be readjusted by using some item of linear detail, which falls within the area causing the discomfort (see "Transference of Principal Points", stage (5)).

With photography of mountainous country, it may prove desirable to re-position the photographs two or three times for each overlap. It is therefore more convenient to hold the photographs in position with weights, than to fix them with cellulose tape or drawing pins.

ADJUSTMENT OF STEREOSCOPES

In a correctly adjusted mirror stereoscope, all the optical paths shown in Fig. 14 should be in the same plane.

To test that a mirror stereoscope is correctly adjusted, put a straight edge under the stereoscope approximately parallel to the axis, and view under high power. In general, two approximately parallel, ghost-like, straight edges will be seen. Rotate the stereoscope slightly to bring the two ghost-like images into coincidence. If a single straight edge is then seen, the stereoscope is in adjustment; but if two ghost-like straight edges are seen crossing one another at a small angle, then the positions of the mirrors require adjustment.

CHAPTER 4

Interpretation—General

THE information on a single, black and white photograph is recorded solely by variations in tone; in spite of this, the viewer normally has no difficulty in interpreting it, if he is familiar with the objects photographed and the relationship, which is remote, that exists between photographic appearance and reality. In other words, the viewer can interpret single photographs reliably, providing he has learnt to do so by his past experience.

When stereopairs of aerial photographs are viewed, an extra interpretative criterion becomes available; this is the relief of the stereomodel. Many more data can be presented to the viewer by a combination of relief and tonal variations, than by either of these criteria separately; thus the stereoscopic interpretations of stereopairs are more reliable than the interpretations of single photographs.

Field and photogeological mapping differ in that the latter has two stages of interpretation, whereas the former has only one. The photogeologist first has to interpret photographic data, which consist merely of variations of tone and relief on the stereomodel, in terms of landforms, man-made lines, structurally controlled lines, vegetation, rock outcrop boundaries, rock outcrop textures, and so on. These constitute his photogeological data. The second stage of his work consists of interpreting these photogeological data in terms of geological structures and lithology. Because both stages are carried out almost simultaneously, the fact that there are two stages is usually overlooked.

The geologist in the field sees his landforms and outcrops directly; he thus has only one stage of interpretation.

RELIEF AND TONE

Relief and tone on aerial photographs are not absolute quantities for particular rocks.

A rock that forms a positive morphological feature when flanked by one type of country rock may form a negative feature when flanked by another. This is illustrated in Plate XIIA, in which the dolerite dyke forms a negative feature between points 3 and 4 in the sandstone but a positive feature between points 1 and 2 in the (?) mudstone. Thus relief on a stereomodel indicates not the absolute resistance to erosion of a rock but its relative resistance compared with the flanking rocks.

Similarly, the tone of a photograph may be affected by many factors, such as nature of rock photographed, light conditions at time of photographing, characteristics of film, influence of filters, effects of processing, and so on.

The same rock outcrop may exhibit different tones on two adjacent photographs of the same strip, because a different angle was subtended by the sun and the aerial camera at the rock outcrop, at the time of exposure, and this resulted in a difference in the amount of reflected light entering the camera. Even from the same negative, different tones may be produced by different methods of printing. Thus only limited use can be made of the absolute tonal value of a photographic image of a rock outcrop for lithological interpretation; it is the relative tonal values that are more useful.

However, on any particular print, the relative tonal values may be very significant. By considering only relative tones, most of the non-lithological factors producing variation in tone are removed. It becomes possible to make certain generalizations that are of value in lithological interpretation. These generalizations refer only to the relative tonal values of the photographic images of rock outcrops on individual photographs; they apply, with reduced reliability, to rocks with residual superficial cover; they do not apply to vegetation-covered outcrops.

Subject to the qualifications above, it may be noted that:

(1) The tone of the photographic image of an intrusion is related to its composition. The more basic intrusions produce the darker tones.

(2) Among the bedded rocks, chalk, limestone, sandstone, quartz-schist, and quartzite tend to photograph in a light tone, mudstones, shales, slates and micaceous schists in intermediate tones, and amphibolites in dark tones.

In the complex igneous intrusion depicted in Plate XVA, area 1 is peridotite, area 2 is gabbro, and area 5 is monzonite. It will be noticed that the peridotite is represented by a darker tone than the gabbro, which in its turn appears darker than the monzonite; this illustrates generalization (1) above. In other words, the tonal values in the different areas in this illustration are related to the composition of the rocks outcropping in these areas; the more basic the outcropping rock, the darker is the photographic tone. It is unfortunate that such a relationship cannot be relied upon in comparing tones between different photographs.

An excellent tonal contrast can be seen in Plate XVB between the dark-toned basalt 3 and the light-toned limestone 5.

Photogeological interpretation is ultimately dependent upon relative tone as estimated on a particular stereomodel, and morphological expression (henceforth called relief), which is a measure of the relative resistance to erosion of the different rocks.

FACTORS AFFECTING THE PHOTOGRAPHIC APPEARANCE OF ROCKS

The factors which affect the photographic appearance of a rock are: (a) climate, (b) vegetation cover, (c) soil cover, (d) absolute rate of erosion, (e) relative rate of erosion of the rock compared with that of the surrounding, country rock, (f) colour and reflectivity, (g) composition, (h) physical characteristics, (i) depth of weathering, (j) structure, (k) texture, and (l) factors inherent in the type of photography and the conditions under which the photography was obtained.

It will be seen that not only are there a large number of factors in the list above, but that many of them are interrelated. The absolute rate of erosion affects the vegetation cover, the soil cover, and the depth of weathering; it is itself affected by the climate, the vegetation cover, the soil cover, the mineral constituents, the physical characteristics, the depth of weathering, and the texture. It is recommended that the reader consider the factors in the list above, one at a time, and decide which other factors they affect and which other factors affect them.

The large number of interrelated factors that affect the photographic appearance of rocks results in this appearance being very variable (Allum, 1961).

LINEAMENTS

In the course of the discussion of photogeological interpretation, we shall make use of the word **lineament.** A lineament is any line on an aerial photograph that is structurally controlled; it includes, for photogeological purposes, any alignment of separate photographic images such as stream beds, trees, or bushes that are so controlled.

The word thus has very wide applications; it can be used to refer to lines representing beds, lithological horizons, mineral banding, veins, faults, joints, unconformities, and rock boundaries. Many examples of such lineaments are illustrated on Plates III–XVI.

Many lines, such as those representing railways, roads, paths, animal tracks, and field boundaries, will be seen on aerial photographs, which although perhaps influenced by the underlying geological structure, are not controlled by it. Such lines should not be referred to as lineaments; thus line 12 in Plate XI, which represents a path, is not a lineament.

LEARNING TO INTERPRET

The best way a geologist can improve his interpretative ability is by working in the field with the aerial photographs and making repeated comparisons of the stereomodel with the ground view. These comparisons are effective both ways. When the geologist compares a particular feature on the stereomodel with the corresponding ground view, he is providing himself with the essential link between the stereomodel and the earth's surface, which enables him to interpret the stereomodel in terms of photogeological data (see beginning of this chapter). When he makes observations on the ground and then studies the corresponding parts of the stereomodel to see how they are represented, he gains knowledge as to the significance to be attached to minute details on the stereomodel.

Another way to improve interpretative ability is to compare good quality geological maps with the relevant aerial photographs. The work with the photographs must be done stereoscopically, with great care, and at all times the position on the stereomodel must be precisely related to that on the map. In general, it will be found that this method is of value only if the geological map satisfies the following conditions:

(1) The scale is not smaller than about 1 : 100,000.
(2) The drainage is depicted in detail.
(3) That it has been mapped in detail or alternatively incorporates the information from a detailed, expert, photogeological interpretation.

Requirements (1) and (2) are obviously necessary, if the photographs are to be correctly related to the map; requirement (3) may cause surprise, however. If the geological map is of ordinary, regional, reconnaissance standard and does not incorporate a photogeological interpretation, experience shows that many vital structures may be unrecorded. The student comparing such a map with the photographs, will begin by losing confidence in himself and photogeology, but will end by developing an unwarranted and unflattering opinion of the geologist responsible for the map.

Theoretically, it would be better to compare the photographs with a high quality, detailed map made without the assistance of photographs, thus avoiding the comparison of one interpretation with another. Such maps are rare, however, and the student must normally be prepared to compare the photographs with a map that represents a synthesis of field and photogeological data.

PHOTOGEOLOGICAL SYMBOLS

It is in the interest of the user of a geological map that he should know the source of the geological data. Therefore, if structures that have been interpreted from photogeological evidence are to be recorded on the final geological map, they should be indicated by the use of special symbols.

In Table 1, a list is given of the more important photogeological symbols. When used in conjunction with field symbols, it may require amendments or additions. The geologist should then invent his own symbols, in general conformity with those listed, and ensure both that their meaning is recorded on the map, and that they are easily distinguishable from the corresponding field symbols.

The latter undertaking is quite straightforward when symbols are represented in different colours, but it is more difficult when only one colour can be employed. In these circumstances, it may be desirable to accept a compromise for certain structural symbols such as those representing faults

TABLE 1

Symbol	Meaning	Colour if used
	Bedding scarps with dip slopes	
	< 10°	Purple
	10° - 25°	Purple
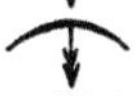	25° - 45°	Purple
	> 45°< 90°	Purple
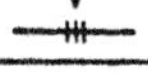	vertical	Purple
	Bedding traces dip slopes absent or very short	
	< 10°	Purple
	10° - 25°	Purple
	25° - 45°	Purple
	> 45°< 90°	Purple
	vertical	Purple
	Horizontal bedding	Purple
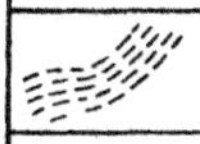	Foliation	Purple
	Lithological or formation boundary	
•••••••••••	Certain	Purple
••• ••• •••	Uncertain	Purple

Table 1 (continued)

	Joints	
	If in colour, distinguishing symbol not required.	
J	Certain	Blue
J	Uncertain	Blue
	Faults	
	Downthrow side, if known, indicated by tick.	
	Certain	Red
	Uncertain	Red
	Tear fault	Red
	Lineament of unknown origin	Purple
	Folds	
	Axial traces with direction of plunge	
	Antiformal	
	Certain	Red
	Uncertain	Red
	Synformal	
	Certain	Green
	Uncertain	Green
	Dykes	
D	Certain	Red
D	Uncertain	Red

or axial plane traces. These should in any case be differentiated into certain, probable, and possible, whether they are based on field or photogeological data; once this is done, there is perhaps no necessity to indicate specifically the source of the data that gave rise to the interpretation. So long as the photogeological symbols representing bedding, foliation, and formational and lithological boundaries are clearly differentiated from the corresponding field symbols, there can be little doubt in the map user's mind as to the relative amounts of field and photogeological evidence that have given rise to a particular structural interpretation.

A source of misunderstanding can arise from the habit of referring to the bedding lineaments seen on photographs as "strike". They represent in fact the trace of the bedding on the surface of the ground and will henceforth be referred to as bedding traces or bedding lineaments.

The processes of erosion of many sediments and metasediments produce scarps and dip slopes; in these circumstances, if the dip arrows are accurately drawn the full length of the dip slopes, as in Plate XIIA, and Plate VII near point 5, the bedding scarp symbols in Table 1 give an excellent indication of the structure. In many sedimentary areas and in most areas of metasediments, however, the processes of erosion do not produce dip slopes of sufficient length for this symbol to be used. In these areas, the bedding trace symbols should be used with ticks to indicate the general direction and approximate amount of dip, as drawn in Plate VIB, near point 9.

There is an obvious source of ambiguity between a short bedding trace symbol and a long foliation symbol. This is deliberate and reflects the likely ambiguity of the interpretation. The longer, more clearly defined, and more persistent the lineaments are, the more likely they are to represent relic bedding rather than foliation. If therefore the symbols are drawn so that they extend the full length of the lineaments, their length will indicate to the map user the likelihood of their representing bedding.

THE STATUS OF PHOTOGEOLOGY

The aerial photograph, like a thin rock-section, should be regarded as a source of geological data. The petrographer, using his microscope, studies a very small piece of rock very closely; all those characteristics of the rock that require an area larger than that of a thin section to become apparent are unseen by him. The field geologist has not the advantage of seeing the microscopic characteristics of the rock; but he sees all the macroscopic characteristics that are able to express themselves in the area of a single outcrop. The photogeologist, in comparison with the field geologist, is still further removed from the rock. Many of the macroscopic characteristics, expressed in single outcrops, are lost to him, but in partial compensation, he "sees" those structures and rock characteristics that require great areas for their expression.

His interpretation of the photographs reflects his geological training, his field experience with aerial photographs, and his general geological knowledge and ability. This is the reason why it is impossible for anybody who is not a geologist to produce a useful photogeological interpretation.

There are certain prevalent, erroneous opinions about photogeology: one is that photogeology is in some way an alternative to field geology; another is that every structure interpreted from photogeological evidence must necessarily be seen in the field before being plotted on a map. Both these attitudes produce unfortunate consequences. The first results in disillusionment if discrepancies are noticed between the interpretation and later field observations. It is then thought that photogeology is unreliable. In fact, of course, the data provided by the photographs are correct, but these, like those from any other source, may be incorrectly interpreted or may be insufficient to ensure a unique interpretation. The second attitude is absurd and inconsistent; it can result in the geologist failing to record structures on his map for which the photogeological evidence is conclusive. It is in those areas where field observations are inconclusive that the geologist should pay increased attention to the evidence provided by the aerial photographs. The less well-exposed an area is, the more important is the information obtained from such sources as photogeology and geophysics.

If a particular rock or structure is seen in the field, that one observation is conclusive. In many complex or ill-exposed areas, however, structures have to be mapped on the evidence of many separate field observations, without the structures themselves ever being seen. These mapped structures are interpretations, similar in nature and significance to those based on data from other sources, such as the thin rock-section or aerial photograph. It is quite wrong for the geologist, merely as a matter of principle, to attach greater importance to one interpretation than another. Under some conditions, the interpretation of the field data might be the more reliable, but under other conditions, this may be true of the photogeological interpretation.

The area represented in Plate IVA illustrates these latter conditions. One sedimentary horizon is indicated by points 8, 7, 6, and 11, and another, by points 4, 5, 2, and 3. It will be seen that portions of the horizon have suffered the following displacements: 7 to south-east relative to 8; 6 to north-west relative to 7; and 11 to south-east relative to 6. Similarly, with the other horizon, the displacements are: 5 to south-east relative to 4; 2 to north-west relative to 5, and 3 to south-east relative to 2.

Considering, for the moment, only portions of horizons 8, 7, 4, and 5, it can be seen that 7 has been displaced relative to 8, in the same sense, and approximately the same amount, as 5 has moved relative to 4. The line, 1–12, along which this movement has taken place is quite clearly defined. Similarly, if portions of horizons 7, 6, 11, 5, 2, and 3 are considered, it will be seen that there are two more parallel lines of displacement marked 9 and 10. All three lines of displacement are fault lineaments; and there are several others on the illustration.

The photogeological interpretation of these lineaments is of a very high order of reliability both as to their origin and location. The reason for the high reliability is the relatively large area which is "seen" simultaneously on the stereomodel. The displacement of 7 relative to 8 would suggest the possibility of a fault, whether it was seen in the field or on the photographs; but seeing simultaneously the displacement of 7 relative to 8, and 5 relative to 4, along a common lineament 1–12, makes the photogeological interpretation as conclusive as anything ever is in geology. The many separate items of photogeological data, all indicating the presence of the same structure, give the interpretation of this structure a high order of reliability. This is known as the **principle of convergence of evidence.**

As a matter of fact, the interpretation of faults in Karroo rocks, such as those in Plate IVA, from field evidence alone, is far more difficult than the appearance on the photographs would lead one to expect. Because of the friable nature of the rocks, fault planes are seldom seen, and outcrops are rare away from the river channels. Although the change in vegetation would suggest a possible change of lithology to the geologist in the field, it should be remembered that this vegetation change is more obvious and more abrupt on the photographs, because of the smaller scale at which it is "seen". Furthermore, the changes of relief, which reflect the changes in lithology in the area illustrated, are greater on the stereomodel than in the

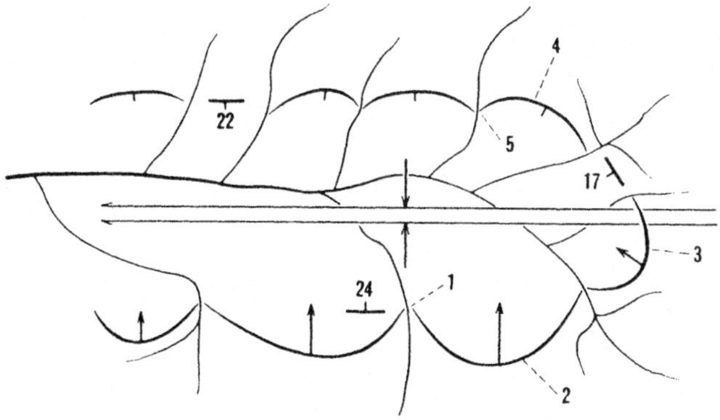

FIG. 15.

field, because of vertical exaggeration (Chapter 3). Thus the photogeological interpretation of the area depicted in Plate IVA is likely to be structurally more reliable than a geological map of the same area based entirely on field evidence.

The geological data obtained from aerial photographs must be regarded as having the same status as those from any other source; and the final geological map should be a synthesis of all data from all sources. To achieve this purpose, it may prove desirable to use both photogeological and field symbols on the final geological map. There should be no objection to this, because the geological picture conveyed by a combination of field and photogeological symbols may be much more comprehensible than that using field symbols alone. The synclinal structure depicted in Fig. 15 is much clearer than it would have been if only the three field symbols had been used.

CHAPTER 5

Interpretation—Structural

BEDDING

Dipping, heterogeneous sediments or metasediments are commonly depicted on a stereomodel as a number of parallel ridges and valleys (see Plates VIB and VII). The ridges and valleys represent the effect of differential erosion on dissimilar rocks. Rocks which differ in their mineral constituents differ in their erosional qualities; resistant beds tend to form ridges along the length of their outcrops, whereas the more easily eroded beds tend to form valleys. Because of vertical exaggeration (see Chapter 3), these ridges are more clearly defined on the equivalent stereomodel than on the ground; they may even be visible on the stereomodel without being visible on the ground. Thus aerial photographs provide evidence of bedding which may not be available from field observations.

Beds which differ in their mineral constituents will also normally differ in their colour and reflectivity. This difference, unless obscured by soil or vegetation cover, is represented on the stereomodel as a difference in tone (see, however, "Relief and Tone", Chapter 4). In Plate VIB, the tonal difference between ridges 8 and 6 is very obvious, and this difference confirms the interpretation that each ridge represents a distinct bed.

The area depicted in Plate VIB is desert with little soil or vegetation cover; the variation in photographic tone of the rock is thus available as an interpretative criterion. Experience in interpreting such areas is valuable because of the confidence that is gained as to the significance of relief which, in areas of soil or vegetation cover, necessarily becomes the dominant interpretative criterion.

The lineaments resulting from a number of conformable, dipping, heterogeneous sediments or metasediments can be expected to show the following characteristics:

(1) They should be persistent. If a bed forms a horizon that is more resistant to erosion than the adjacent country rock, it will form a ridge; the ridge may be interrupted by streams following planes of

weakness such as joints, but the whole lineament, which consists of all the short lengths of collinear ridges, should persist throughout the full length of the outcrop, see Plate IX, ridge 1.

(2) They should be approximately parallel to one another. The abrupt cessation of this parallelism provides evidence for other structures.

(3) They tend to be found in groups rather than singly.

(4) They tend to have a definite width and to be limited in number; this distinguishes them from lineaments resulting from foliation, which appear indefinitely thin and unlimited in number.

In practice, all the characteristics are observed almost simultaneously and the interpretation is made according to the convergence of evidence principle (see Chapter 4).

It is unlikely that any particular example of dipping sediments will exactly fit the list of characteristics given above. For example, in Plate VII, the ridges are clearly defined, but ridge 3 does not represent a single bed but rather a thin group of resistant beds; ridge 5, however, appears to be caused by a single bed. It is worth studying ridge 3 with some care in order to decide to what extent it is reasonable to rely on such ridges to indicate the bedding structures. Ridges 5 and 2 give a good representation of the folded bedding, but ridge 3 indicates only an approximate direction of strike.

For horizontally bedded sediments, or those of low dip, the approach is necessarily very different. The shapes of the outcrops approximate to the shapes of the contours. If a horizontal bed is particularly resistant to erosion, it does not produce a ridge but merely a steepening of the ground slope where it outcrops (e.g. in the sides of river valleys).

An example of almost horizontal bedding is given in Plate IIIB. The changes of slope are very obvious between points 6 and 7 and also between 2 and 1. Each change of slope indicates a bed; confirmatory evidence for this is provided by the distribution of vegetation, which grows preferentially on certain beds.

Beds 8 and 5 have been completely isolated by erosion. At points 3 and 4 it is possible to compare the photographic appearance of a dipping bed with that of the same bed when almost horizontal.

This example is exceptionally easy to understand because it depicts a desert environment. Under moister conditions, the evidence would probably be of the same type but would be less convincing.

A more difficult stereopair to interpret is Plate IVB. The beds are almost horizontal, and are indicated by change of slope and vegetation between points 8–7 and 4–3. The evidence is similar in nature to that of the previous example but less convincing.

The evidence at points 1 and 2 on the western edge of the stereopair is more conclusive and proves that the rocks consist of almost horizontally bedded sediments; but even in the eastern sector there is enough evidence to suggest this interpretation. The various flat surfaces, such as those at points 6 and 5 and the flat-bottomed valley 9, indicate that the rock is layered horizontally.

Another example of horizontally bedded rocks is represented in Plate V, point 14. Once again the evidence consists of changes of slope and distribution of vegetation.

The above discussion of bedding applies to metamorphosed as well as unmetamorphosed sediments; in fact, the ridges 8 and 6, in Plate VIB, are both metamorphosed and represent limestone and quartzite respectively. Metamorphism does, however, complicate interpretation.

At this stage in the discussion, the concept of the **constant lithological horizon** becomes of value. Strictly speaking, the term "horizon" normally refers to the surface between two adjacent beds, rather than to a particular bed; but for photogeological purposes, it is convenient to be able to refer to bands of rock of constant lithology as constant lithological horizons; and we shall use the term in this sense. It is analogous to a geologist referring to a bed of distinctive lithology as a **marker bed** or a **marker horizon.**

In the discussion of photogeology, it is always necessary to remember that the photogeological data indicate lithological horizons, and that the question as to what extent the lithological horizons represent bedding, is a separate problem. In unmetamorphosed areas a large number of consistently parallel lithological horizons may be interpreted as beds with a high degree of confidence. Even in these areas, however, some of the horizons could represent conformable lava flows or sills. In metamorphosed areas, the possibility of error in interpreting lithological horizons as beds is greater.

In many areas of metasediments, the relevant stereomodel still shows subdued and obscure ridges even though there may be virtually no field evidence of bedding. For example, the rocks represented in Plate VIII are phyllitic. The question arises as to what significance to attach to the subdued ridges such as 1–2, 3–4, 6–9, and 7–8. Although from field work the foliation of the phyllitic rocks is known to be approximately parallel to the ridges, it is not believed that it caused them (see section on "Foliation"). Over most of the area, there is little evidence of bedding to be seen in the field but zone 5, enclosed by dotted lines, was found to contain pebbles. If the pebbly zone indicates the bedding, as is probable, then the parallelism between the pebbly zone and the subdued ridges suggests that the ridges may also represent bedding and may be caused by differences in lithology not apparent in the field.

The persistence of the particular ridges, which is obvious only on the aerial photographs, is itself evidence (though not conclusive), that the ridges represent individual lithological horizons, and therefore probably beds.

When numerous, parallel, subdued ridges are seen on the stereomodel of an area that from field work is known to consist of apparently homogeneous metasediments, thought should be given as to their significance. They may represent beds that cannot be differentiated by the unaided eye.

DIP

Dip slopes provide the most reliable indication of the direction of dip available to the photogeologist. Dip slopes in sediments are illustrated in Plate VII, and Plate V, point 6, and in metasediments, in Plate VIB, points 6 and 7.

Many dip slopes are more reliably recognized on the stereomodel than in the field; this is because, on the stereomodel, a large area of the dip slope can be "seen" simultaneously and can be compared with other horizons in the same group of sediments.

Although dip slopes on the stereomodel provide useful evidence as to the direction of dip, their indication of the amount must be treated with reserve. The two main reasons for this are: firstly, vertical exaggeration makes all dips appear steeper than they are in reality; and secondly, relief displacement makes all vertical structures appear to dip towards the principal point of the photographs. The relief displacement near the centres of the photographs is much less than towards their periphery; for all estimates and measurements of dips, therefore, it is desirable that only the central portions of the photographs should be used.

Accurate measurements of dip may be made on aerial photographs by photogrammetric methods (see Chapter 10), but in regional mapping, the time available usually precludes this. In such conditions, photogeologists confine themselves to approximate estimates of dip such as horizontal, $< 10°$, 10°–25°, 25°–45°, and $> 45° < 90°$ dips (see Chapter 4, Table 1).

To make the dip estimates, the dips on the stereomodel are first estimated, and then an adjustment is made to allow for vertical exaggeration.

If only the central portions of the photographs are used and if the photography is truly vertical, horizontal beds will appear horizontal on the stereomodel and vertical beds vertical; it is only the moderate and steep dips on the stereomodel that have to be corrected. Thus, apparent dips on the stereomodel of 35° and 60° might be recorded on the interpretation as 10°–25° and 25°–45° dips respectively.

It is possible to develop formulae for calculating vertical exaggeration and also correction charts relating apparent dips on the stereomodel with the true dips in the field. For these, Ray (1960) and Miller (1961) should be consulted.

In many areas, dip slopes are not developed; the dips of the sediments or metasediments may express themselves merely through the shapes of the bedding traces. Subject to the qualification concerning relief displacement near the periphery of the photographs given above, vertical beds produce bedding traces which are independent of the relief of the ground and which therefore do not deviate as they cross hills and valleys. Dipping horizons, however, as they cross interfluves and river valleys, produce crescent and V-shaped traces respectively. In Fig. 15, somewhat idealized crescents and V's are illustrated at points 2, 3, 4, and 1 and 5 respectively.

Actual examples of outcrop traces forming crescents, V's, and dip slopes are represented in Plate VIB, bed 7, and Plate VII, beds 5 and 2. The angle of the V's points in the direction of dip.

In Plate IVA, the only indication of the gentle dip to the north-west is the crescent shape of the bedding traces such as 13. Interpretation of dip in such areas consists of noting the shape of the crescents and the angles of the V's relative to the relief, estimating whether the dip is gentle, moderate, or steep on the stereomodel, and making an arbitrary correction for exaggeration. For any particular river valley, the more obtuse the angle of the V, the steeper is the angle of dip.

FOLIATION

Discussion of foliation is bedevilled by the fact that British and American geologists use the word in different senses. British geologists distinguish between schistosity and foliation; for them, the segregation of platy or elongate minerals into thin layers or folia is called "foliation", whereas the parallel orientation of such minerals is referred to as "schistosity" (Harker, 1950, pp. 194 and 203). American geologists give a wider meaning to "foliation". Turner and Weiss (1963, p. 97) define it as follows: "we use the term foliation to cover all types of mesoscopically recognizable s-surfaces of metamorphic origin. Lithological layering, preferred dimensional orientation of mineral grains, and surfaces of physical discontinuity and fissility resulting from localized slip may all contribute to foliation of different kinds." This latter definition is suitable for photogeology and we shall use the word in this sense.

Lineaments resulting from foliation may be expected to show some or all of the following characteristics:

(1) They should tend to be parallel to one another. Because regional foliation results from regional stress, the foliation planes should be parallel to one another over a considerable area unless they have been distorted or folded by subsequent movements. Some areas, however, subjected to more than one period of metamorphism, exhibit more than one set of parallel foliation planes.

(2) They should be very numerous. Because the number of foliation

planes in the rocks is very large, the number of lineaments one can expect to find on the photographs is also very large.

(3) They should be short. Because one foliation plane is similar to another, there is no property that can ensure the continuation of a particular lineament. For example, if a stream flows parallel to the foliation and reaches a line of greater weakness such as a transverse joint, it should continue in the transverse direction to the end of the joint, and then take up its original direction parallel to the foliation; its final course, although parallel to its original one, should not be in the direct prolongation of it, and there should thus be two short, parallel, separate, linear features. In contrast to this, if the stream had been in a bed of metasediments appreciably easier to erode than the surrounding rocks, it would tend always to return to the soft bed, or alternatively a new stream would develop in the soft bed; whichever event occurred, the soft bed would form a long, single, and almost continuous feature.

(4) They should not consist of long continuous ridges or valleys for the reasons given under (3) above.

Many ridges seen on the stereomodel and believed to represent distinct, metamorphosed beds, are covered with innumerable short lineaments parallel to the foliation seen in the field. Lineaments of this kind, in this case represented by the alignments of trees and bushes, can perhaps just be seen in Plate VIII, on the top of the ridge 7–8*. It is believed, by the author, that the major lineaments (ridges) represent the beds and the minor lineaments the foliation (see characteristics of bedding and foliation lineaments listed above).

In a rock that has vertical or steeply dipping foliation but is otherwise homogeneous, erosion tends to produce lens-shaped outcrops, in which the long axis of the lens is parallel to the foliation seen in the field. This characteristic can be very useful in determining the foliation direction in granitic areas from aerial photographs.

FOLDS

Where the photographs indicate the direction and amount of dip of the beds, as in Plate VII, beds 5 and 2, it is possible, from photogeological evidence alone, to map the approximate position of the axial trace (i.e. the trace of the axial plane on the surface of the ground) and to estimate the direction and amount of plunge of the fold. In this example, the axial trace appears to be curved and to pass through bed 5 in a north-north-west direction; the approximate direction of plunge is north-north-west.

*They are not clear on the half-tone reproduction. There is always some loss of detail when photographs are reproduced, and it is not possible to be certain beforehand which details these will be.

Photogeological estimations of the positions of axial traces and the directions and amounts of plunge are quick to make, and valuable. The geologist sees on the stereomodel a representation of the whole fold, in three dimensions, simultaneously; his estimation of the axial trace position and plunge may be more reliable than similar estimations based on dips and strikes plotted on a geological map or "seen" separately on aerial photographs. This does not mean, however, that such photogeological estimations are comparable in reliability to the results of structural analysis based on detailed field work.

A beautiful antiformal structure is illustrated in Plate XIIIB. (The structure has not been visited in the field, so the statement has not the support of field evidence; but does it need it?) Ridges 9 and 10 represent beds. If they are studied carefully, it will be seen that the beds dip away from one another. Where the two beds come together at 6, a fold nose is formed. The imaginary line joining points 6 and 2 represents the plunge of the fold.

Note how the two bedding lineaments 9 and 10 combine to produce one bedding lineament 2; this is indicative of gently plunging folds. The length of lineament 2 depends on the plunge of the fold and the slope of the ground. In the area depicted, lineament 2 soon terminates as the plunge of the fold takes it below the surface of the ground; but if the plunge had approximated to the ground-surface slope, the lineament would have been longer.

It is worth while studying this stereopair and imagining the effect of varying the angle of plunge. If the fold axis were vertical, the fold arms 9 and 10 would form a fold nose but lineament 2 would not be present. If the axis were parallel to the earth's surface, the fold arms 9 and 10 would be parallel to one another. Such different patterns of bedding traces are numerous in soil and vegetation-covered areas of sediments and metasediments.

The area depicted in Plate XIIIA consists of phyllitic rocks, similar to those represented in Plate VIII. In the discussion of Plate VIII (see sections on "Bedding" and "Foliation"), reasons were given for regarding the ridges as representing beds rather than foliation. If this reasoning is accepted, then the ridges depicted in Plate XIIIA, which have the characteristics of lineaments resulting from bedding, may be interpreted as beds. This constitutes an important advance, because the field geologist is often able to observe only foliation in such areas and it is the attitude of the beds that is of the greater interest.

The tentative photogeological interpretation of Plate XIIIA should be that of folded beds with each ridge representing one bed.

In Plate IX, the fold structure is indicated by the alternating ridges and valleys 1, 2, 3 and 4; the ridges consist of hornblende gneiss and the valleys of dolomitic limestone. Within area 5, which has all been mapped as hornblende gneiss, ridges can be seen, which indicate the same fold structure.

This is important, because it shows the significance of ridges, even in apparently lithologically homogenous rock masses.

After studying Plate IX, it becomes possible to interpret with some confidence the ridges in the vicinity of point 1 in Plate XIVA as representing a fold nose.

The aerial photographs of some areas indicate structures that would certainly be unnoticed in the field. The area depicted in Plate X is flat and soil-covered, yet the fold structures are clearly indicated by the distribution of small trees and bushes.

By plotting bedding traces on aerial photographs, it is possible in places to prove the presence of repeated folding.

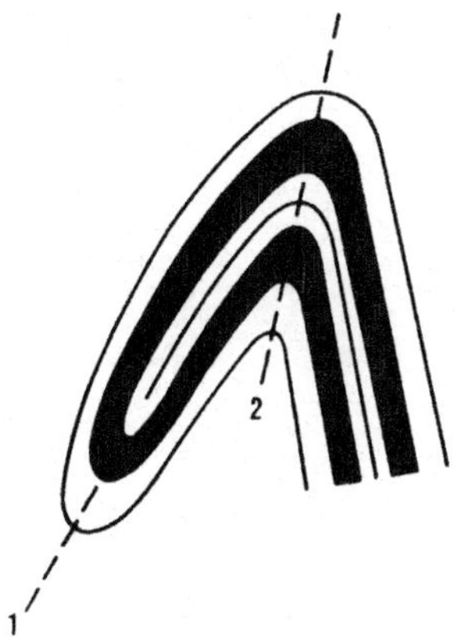

FIG. 16.

If the bedding traces have the form shown in Fig. 16, the structure must have been folded at least twice; the axial trace of the first formed fold being indicated by 1 and that of the second by 2. It is probable that Plate X represents such a structure, with the line 2–5 representing the axial trace of the first formed fold and 1–3 representing that of the second. The whole structure extends over many photographs and is too large to reproduce in a book.

Subsidiary folds can be faintly discerned just north of point 4.

FAULTS

A fault may be defined as a fracture along which there has been slipping of the contiguous masses against one another (Lahee, 1941). Characteristically they form more or less straight, negative features in the earth's crust. There are, however, many exceptions and qualifications to this generalization.

Many fault planes become injected with magmatic or mineralizing solutions to form dykes or mineral veins; they are then likely to form positive rather than negative features. Furthermore, the photographs indicate the trace of the fault plane on the surface of the ground; unless, therefore, the fault plane is vertical, its trace will be affected by relief.

Before a fault can be interpreted, there should be some evidence of movement; in most areas, this evidence consists of the termination or displacement of other structures. Displacement is a more reliable criterion than termination, because structures can be terminated, otherwise than by faulting. Bedding gives rise to more reliable evidence than other structures, because

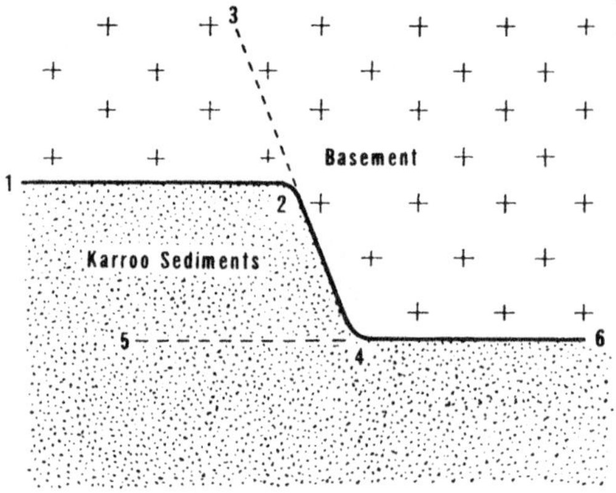

FIG. 17.

it is more predictable. Thus the most reliable evidence of faulting is the displacement of bedding along negative linear features. Such evidence is illustrated in Plate IVA, points 9, 10, and 12; it was discussed in Chapter 4 (section "The Status of Photogeology").

The displacement or termination of other structures, such as dykes, other faults, and intrusive contacts, all provide evidence of faulting; but this evidence is of a lower standard of reliability.

When the interpretation of a major lineament is in doubt, it is advisable to study all the structures on both sides of the lineament with care, to see whether all the evidence, considered simultaneously, indicates a fault. It is not necessary for any particular piece of evidence to be almost conclusive, for the final interpretation to be so. If several items of evidence independently suggest that a particular lineament represents a fault, then it is almost certainly one (principle of convergence of evidence).

The photogeological evidence for the faults forming the boundaries of rift valleys is, in many places, of this kind. Consider the faults illustrated diagrammatically in plan in Fig. 17. The scarp forming the side of the valley lies along the lines 1–2, 2–4, and 4–6. The Basement rocks are at a higher elevation than the Karroo sediments, which cover the bottom of the valley. A lineament can be seen from point 2–3; and a faint line is indicated by vegetation, in the valley bottom, from point 4–5.

The photogeological evidence for the faulting is stronger than is usually realized and consists of:

(1) The difference in elevation between the Basement and Karroo rocks.
(2) The straightness of the scarp line 1–2.
(3) The straightness of the scarp line 2–4.
(4) The straightness of the scarp line 4–6.
(5) The fact that only one of the lines 1–2 and 4–2 continues into the Basement, i.e. line 4–2 terminates line 1–2 but itself continues into the Basement along line 2–3.
(6) The fact that only one of the lines 6–4 and 2–4 continues into the sediments, i.e. line 6–4 terminates line 2–4 but itself continues into the sediments along line 4–5.

All these items of evidence are inconclusive in themselves, but considered together, they lead to an interpretation which has a very high degree of reliability.

The photogeological evidence alone is sufficient to prove many faults, in the area depicted in Plate V. The lineament 1–7, bifurcates at 7, into the two lineaments 7–17 and 7–13. Beds 4 terminate at the lineament 1–7, and beds 11 and 15 terminate at lineaments 7–13 and 7–17 respectively. The three terminations of the bedding leave no doubt that the three related lineaments are faults.

Beds 9 and 16 resemble each other; the photogeologist, therefore, naturally studies the area around 9 and 16 to see whether there is any evidence that the beds are identical and that 9 had been faulted down, relative to 16. In this illustration, comparison of the beds underlying 9 and 16 suggests their lithological similarity, but does not prove them to be the same beds.

Careful photogeological study of the area between 9 and 16 indicates that there is a structural break between them. A lineament can be traced from 18, via point 12 to point 10. This lineament terminates bedding at point 12, and it therefore probably represents a fault trace. The beds 6 terminate against bed 9 along a lineament 8–3, which thus also probably represents the trace of a fault. (The trace is somewhat obscured because material from bed 9 appears, in the vicinity of point 8, to have washed partly down the dip slopes of beds 6.) The continuation of lineaments 3–8, in a southerly direction forms the western termination of bed 9 and appears to join lineament 18–10. There is, therefore, a high degree of probability that the whole lineament 18–3 represents a fault trace. At point 3, lineament 18–3 meets another lineament 2–5, which not only terminates it but also forms the northern boundary of bed 9; lineament 2–5 therefore also represents a fault. Supporting evidence for regarding lineament 2–5 as a fault can be found by comparing the photographic appearances of the beds on either side of it. (There is some photogeological evidence that fault 2–5 consists of two separate but almost collinear faults.)

This faulting has been discussed in some detail in order to illustrate the photogeological approach to structural problems. The photogeological data obtained from the photographs have to be correlated and built into a single picture. Once one structure has been proved, it can be used to provide evidence for other structures.

Other faults which have not been discussed are illustrated in Plate V.

The presence of sediments surrounded by Basement, as depicted in Plate VII, suggests that the sediments may have been preserved by down-faulting. The termination of the bedding at 7 suggests the position of the fault. Taking 7 as our starting point, we find a lineament through the points 6, 7, 8, 9, 10, 11, and 14. The photogeological evidence that this lineament represents a fault is as follows: (a) the lithological horizons 13 terminate at 14; (b) the rocks on either side of the lineament between points 14 and 11 have a different tone; (c) between points 10 and 11 the river follows the lineament; (d) there is a difference in ground elevation between different sides of the river from 10–11; (e) between points 8 and 9 and points 11 and 14 the lineament is a clearly defined negative feature; and (f) between points 6 and 7, the lineament terminates many of the individual sedimentary beds. The evidence for the fault is thus overwhelming and entirely photogeological.

There is no evidence that the lineament 4, which represents the other contact between the sediments and the Basement, is also a fault. This lineament presumably represents an unconformity.

In areas of soil and vegetation cover, faults may be less obvious but the final interpretation may not be less conclusive. The conclusiveness depends upon the number of pieces of evidence which support the same interpretation (convergence of evidence). For example, the area depicted in Plate XIVA is covered with soil and vegetation, yet the photogeological evidence that the lineament 3–4 represents a fault is almost conclusive. The evidence is: (a) the termination of the ridge 2; (b) the difference in mean elevation on either side of the lineament; (c) the difference in vegetation on either side of the lineament; (d) the abrupt change of strike that occurs at the lineament (this only applies if the reader accepts that ridges 2 and 6 represent the strike trace of their respective rock groups). Owing to the limited size of illustration that is possible to publish in a book, only a short length of the lineament can be illustrated. The lineament extends over many miles, but the photogeological evidence that it represents a fault is all of the kind illustrated.

In Plate XIVB, the interpretation of the lineament 4–8 as representing a fault depends upon the recognition that the contact between rocks 7 and 9 is represented by the lineaments 10–6 and 4–2. Once the contact has been recognized, its displacement provides evidence for the fault.

JOINTS

Joints are characteristically represented on stereomodels as straight, negative features; they are thus similar in photographic appearance to faults. Before a fault can be interpreted, however, there should be evidence of relative movement between different sides of the lineament of the kind described in the previous section. If there is no such evidence of movement either direct or indirect, the lineament should be interpreted as representing a joint. It is better to record a joint on the interpretation, which subsequent field work shows to be a fault, than vice versa. The plotting of a joint with a photogeological symbol records the presence of the earth fracture, and draws the attention of the field geologist to it during any subsequent investigation.

The jointing in the sandstone depicted in Plate IVB is very obvious; but in other areas it is not so easy to recognize. The lineaments 5 and 9 in Plate VIB are probably joint controlled. If a stream course is straight and transverse to the strike of underlying sediments, then it is probably controlled by a plane of weakness common to all the beds. If there is no evidence of faulting, then such planes of weakness should be interpreted as joints.

In Plate XI, two directions of jointing are indicated; one, represented by lineaments 5 and 9, is perpendicular to the regional strike trace of bedding and foliation; the other, represented by lineaments 8 and 10, is oblique to it. The joints of the first set, although not well represented on the stereopair, are very well developed in the vicinity of the area illustrated; they are in places more than a mile in length, and, in conjunction with the regional strike, control the drainage. The joint and strike control of the drainage can be seen from the shape of stream-bed 7.

These major joints, perpendicular to the regional strike, are easily overlooked in the field, unless the geologist studies his aerial photographs. Traverses perpendicular to the strike do not cross them. Even stream-bed traverses, which for much of their length would lie along them, would not necessarily reveal them, because in tree-covered country the length of stream-bed that the geologist sees at any one time is so short that he may never learn that it is structurally controlled.

It is probable that, in the past, insufficient attention has been paid to these major joints. They probably indicate the direction of maximum stress of the last period of metamorphism; they should dissuade geologists from postulating that shearing forces, parallel to the regional strike of the foliation, were the cause of the regional foliation.

Lineaments in the granite, of the type marked 13, are not parallel to those of the adjacent country rock 14, and do not displace the granite boundary; they should therefore be interpreted as joints. Support for this interpretation is given by two joint-like lineaments 8 and 10, in the country rock, parallel to lineaments 13.

Joints can be of great assistance in interpretation. The thick, tropical rain-forest cover in Plate XIIB does not hide the rectangular jointing 1 and 2. The jointing, in conjunction with the relief, produces an appearance characteristic of granite within the area; thus it is possible to plot the granite boundary (dotted line) with some confidence and precision.

LITHOLOGICAL BOUNDARIES

In most areas that are being fairly actively eroded, it is possible to differentiate dissimilar rocks on aerial photographs even though it is not possible to decide of what the rocks consist. It has in fact even proved possible, from the evidence provided by aerial photographs, to subdivide rocks which appear indistinguishable in the field.

A photogeological study of the Chambe Plateau, Malawi (Stringer and others, 1956) indicated that it consisted of three concentric ring-dykes around a central plug. The major rock types were all syenites, in which only a few minor megascopic differences were detectable. On the ground, apart from steep outer and inner ramparts, the ring structures were not apparent. An extensive collection of specimens was made and subjected to detailed study. Petrographic criteria were established enabling differentiation to be made between the syenites of the various rings which, in general, confirmed the photogeological interpretation.

Without the photogeological interpretation, the presence of the ring structures might never have been suspected; and even if their presence had been suspected, a very large number of specimens would have been required to establish the distribution of the rings petrographically.

In Plate IX, the boundaries between the limestone horizons, 2 and 4, and the hornblende gneisses, 1, 3 and 5, are clearly defined.

In Plate XIVB, the boundary between phyllite 7 and granulite 9 is fairly obvious; it extends from point 10 to point 6 and point 4 to point 2; less clear is the boundary between the phyllite 7 and the granite 3, which is represented by an indistinct negative feature between points 1 and 5.

Under the desert conditions represented in Plate XVB the boundary between the basaltic lava 3 and the sediments 5 is depicted with the greatest clarity.

Little difficulty should be experienced in plotting the boundary 2 between the granite and the metasediments in Plate VIB. The same type of contact is depicted in Plate VIA, but here both rocks are completely soil-covered. (The contact has been annotated along part of its length.) The two rocks produce different soils and support different vegetation; the differences in soil and vegetation and distribution of vegetation produce differences in photographic tone and texture and leave no doubt as to the position of the contact. In this illustration, the stereoscopic view does not assist in the recognition of the boundary; the boundary may be plotted with greater accuracy by the study of a single photograph. The reasons for this are that there is no difference i· relief on different sides of the boundary and the effect of the tone and textural differences, on which the definition of the boundary depends, is reduced by magnification.

The difference in relief and joint pattern between the granite and the country rock, in the tropical rain-forest area depicted in Plate XIIB, makes their contact, which has been partly annotated, apparent.

The boundaries 4, of an intrusion of monzonite 5 in gabbro, are illustrated and partly annotated in Plate XVA.

In some places two distinct rocks are found, such as a permeation gneiss and an autochthonous granite, without any contact between them; the rocks grade from one into the other. In such areas it is necessary to plot a gradational contact. Because with aerial photographs it is possible to look at a large area and a large number of outcrops simultaneously, it is frequently possible to plot such gradational contacts with greater consistency from photogeological data than from field data alone.

CHAPTER 6

Interpretation—Lithological

IF a geologist has prolonged experience in the field in a particular area with the aerial photographs, it may prove possible for him to recognize the rocks of that area, from the photographs, with considerable reliability.

Normally, however, the expression "lithological interpretation" refers to the recognition of rock types from photogeological data alone, rather than from photogeological data supported by local field experience. It is this branch of photogeology that is discussed here.

The approach recommended is one that combines a geomorphological and structural analysis with the use of a generalized photogeological legend; in this, each outcrop represented on the stereomodel is considered in its local and regional geological environment, and the final lithological interpretation is not made until these other studies are completed. In this chapter, no attempt is made to discuss the probable photographic appearance of the particular rocks; it is assumed that almost any appearance is possible.

Although there is no intention of suggesting a fixed routine of interpretation, the following stages are typical:

(1) The recognition of the climatic environment, e.g. temperate, tropical rain-forest, savannah, desert, etc.

(2) The recognition of the erosional environment, e.g. very active, active, inactive.

(3) The recognition and annotation of the bedding traces of the sediments or metasediments.

(4) The recognition and delineation of areas of outcrop that do not indicate bedding. (These areas possibly represent intrusions, and they are studied for confirmatory evidence; alternatively, they may represent horizontally bedded sediments or metasediments.)

(5) The recognition and delineation of areas of superficial cover that do not indicate bedding.

(6) The restudy of the bedding traces (3) around fold noses to determine, if possible, the approximate position of the axial traces.

(7) The study of lineaments transverse to the bedding traces to determine whether they represent faults, dykes, joints, or combinations of these.

The advantages of this kind of approach are: firstly, it can be used for all rocks under all environments, irrespective of the photogeologist's experience with the particular rocks and environments in question; and secondly, the photogeologist, on making his lithological interpretation by this method, is conscious of the degree of doubt involved.

GENERALIZED PHOTOGEOLOGICAL LEGEND

It is accepted that the field geologist, basing his initial subdivision of the rocks upon his inspection of hand specimens, must use more generalized rock divisions than the petrographer, using data obtained from a microscopic examination of the rocks. Analogously, it is necessary for the photogeologist to use more generalized rock divisions than the field geologist.

Although a satisfactory *key* showing all the possible photographic appearances of particular rocks cannot be produced (Allum, 1961), dissimilar rocks will, in general, appear dissimilar on the photographs; they can be differentiated on photographic evidence alone, although no specific rock names can be given to them. It is, therefore, true to say that a photogeologist may be able to divide the rocks into their respective groups without being able to decide of what the groups consist. It is one of the purposes of the generalized legend to enable these divisions to be recorded without using petrological names that are not justified by the photogeological data. Except in the case of certain sediments that can be recognized specifically, the generalized photogeological legend should indicate the type of rock, rather than the name of the rock; the various groups of metasediments are better differentiated by numbers, rather than by names of doubtful accuracy.

A photogeological legend should be self-explanatory, be not more specific than the photogeological evidence justifies, and should enable all the geological information obtained from the photographs to be recorded.

For a particular body of rock, some or all of the following data may be obtained from aerial photographs:

(1) The photographic tone of the rock body relative to that of adjacent rocks.
(2) The resistance to erosion of the rock body relative to that of the adjacent rocks.
(3) The boundary of the whole rock body.
(4) The topographical expression of the whole rock body.
(5) The boundaries of the individual outcrops.

(6) The joint pattern.
(7) The fault pattern.
(8) The drainage pattern.
(9) The vegetation cover.
(10) The bedding or the relic bedding lineaments.
(11) The foliation lineaments.
(12) The regional geological environment.

It is on these data that the photogeologist depends for the differentiation of his rocks. They should, under suitable conditions, enable him to decide whether a particular rock is of intrusive or sedimentary origin, of acid or basic composition, and also whether it is metamorphosed. They give little indication of the composition of the metamorphic rocks; thus, mineralogical prefixes to the names of the metamorphic rocks are unsuitable for a photogeological legend. The generalized photogeological legend should consist of those rock divisions that can be differentiated, using the data listed above.

An example of the type of generalized legend recommended is given below.

(1) Superficial cover—transported.
(2) Superficial cover—residual.
(3) Sediments—group 1.
(4) Sediments—group 2, etc. (Specific rock names may be used if the rocks can be interpreted with confidence.)
(5) Metasediments—group 1.
(6) Metasediments—group 2, etc.
(7) Metasediments—undifferentiated.
(8) Permeation gneisses and migmatites.
(9) Granitic rocks—autochthonous (mostly granites and granodiorites).
(10) Granitic rocks—intrusive.
(11) Granitic rocks—general (may be autochthonous or intrusive).
(12) Intrusive rocks—basic.
(13) Extrusive rocks—acid.
(14) Extrusive rocks—basic.
(15) Dykes—acid.
(16) Dykes—basic.
(17) Dykes—general (may be acid or basic).

The use of group numbers for distinguishing between rocks of dissimilar photographic appearance, but falling into the same main divisions, can be extended to all the main divisions in the legend, instead of being confined to the sediments and metasediments as in the legend above. The flexibility of this legend is of value because it allows continuous modification and enlargement as the work of interpretation progresses.

It will be noticed that certain quite important divisions, such as the intermediate intrusive and extrusive rocks, are missing from this legend. This is deliberate because the legend is based on the divisions that it may be possible to determine from photogeological data only. If the photogeologist has other data, such as local field knowledge or previous mapping, he may with advantage be able to increase the number of divisions. Whatever data he has, however, he is always likely to find some photographic images that he cannot interpret. These he should delineate and mark with a query.

Some notes on the lithological interpretation of some of the rock divisions of the generalized legend are given below.

SEDIMENTS AND METASEDIMENTS

Sediments are recognized on photographs by their layered appearance. Although the photographic appearance of the sediments represented on Plates IIIB, IVA, IVB, VII, and XVB differ from one another, they all show this diagnostic layering. In general, this layering is represented by variation in both relief and tone, but it may be represented by variation in either of them alone. The beds 5 in Plate XVB are indicated by relief only.

In most areas it is possible to differentiate between metamorphosed and unmetamorphosed sediments. One of the results of metamorphism is to make the individual beds of a pile of sediments more nearly equal in their resistance to the forces of erosion. Dipping metasediments, therefore, tend to form ridges that are more rounded and subdued than those produced by sediments of similar dip. Compare, for example, the ridges formed by metasediments in Plate IX with those formed by sediments in Plates VII and XVB.

Another indication of metamorphism is the geological environment. If the layered rocks are steeply dipping, tightly folded, and are associated with multiple intrusions, metamorphism should be suspected; conversely, flat lying or gently folded layered rocks, unassociated with intrusions, are less likely to be metamorphosed.

Of course, in many places where sediments are laid down unconformably on metamorphosed rocks, both types are found together (Plate VII). In such areas, however, the lithological dissimilarity and structural discontinuity between the two groups make their differentiation obvious.

Even in many soil-covered areas, metasediments may still be correctly interpreted. Plate X probably represents an example of such an area; it is completely soil-covered, yet its interpretation as representing an area of folded metasediments would have a high degree of reliability.

INTRUSIVE ROCKS

Acid

Acid intrusive rocks are typically relatively light-toned and rectangularly jointed. Although in many places they are more resistant to erosion than the country rock and form positive features, no general rule can be made. Greenwood (1962) found that in desert areas, among the plutonic igneous rocks, granite was one of the least resistant to erosion. Allum (1961) noted that two granites, within a few miles of one another, eroded at very different rates.

The boundary between the granite and metasediments in Plate VIB is shown dotted. If this boundary is inspected carefully in the vicinity of point 2, it will be seen that it is apparently not quite conformable with the bedding of the metasediments. Inspection of the stereomodel of the inselberg 3 and the very small outcrops 1 gives no indication of bedding. These observations indicate that the rock south of the boundary is intruded. The light tone of the small outcrops 1, the extreme top of the inselberg 3, and the surrounding superficial cover suggest that the intruded rock is probably acid.

The area depicted to the east of the boundary in Plate VIA is light-toned and almost totally lacking in lineaments. This apparent homogeneity, combined with light tone, suggests an acid intrusion.

In the tropical rain-forest area depicted in Plate XIIB, the north-east area is one of high relief relative to the country rock, and it exhibits rectangular jointing (points 1 and 2); in all other respects, the rock represented appears to be homogeneous. These data, together with the curved nature of the boundary (particularly when seen *in toto*), would enable most photogeologists to interpret the rock correctly as a granite.

In Plate XI, the lithological horizons of the country rock, represented by such lineaments as 4 and 14, diverge in the vicinity of point 6, as though the rock 11 had been wedged or intruded between them. As rock 11 is light-toned, jointed, and otherwise apparently homogeneous, it can be interpreted with confidence as a granitic rock—intrusive.

In the photogeological legend the expression "granitic rocks—intrusive" is used rather than "acid—igneous", because although photogeological evidence is likely to indicate whether a rock body is intrusive or formed *in situ* (autochthonous), it is unlikely to indicate whether it is of magmatic origin. A granitic rock may be formed by processes of granitization, and at a late stage in these processes be intruded in the earth's crust to form para-autochthonous "granite" (Read, 1957). It is not possible, on photogeological evidence, to distinguish between this and an intrusive magmatic granite.

In many areas, the photogeological evidence is insufficient to differentiate between autochthonous and intrusive granitic rocks. In Plate XIIIB, the rock body represented at point 4 can be interpreted as a granitic rock having boundaries indicated by the dotted lines 1 and 3, which can be traced respectively to points 7 and 8. The boundaries can be seen to be conformable to the relic bedding of the adjacent metasediments, but whether this results from the granite having been intruded between the relic bedding planes, or from the preferential granitization of particular lithological horizons, is difficult to determine from photogeological evidence.

A very similar granite is represented between points 5 and 11.

Basic

It is after a rock body has been recognized as intrusive on structural grounds (see Chapter 5) that an attempt is made to decide whether it is acid or basic. If any part of the rock body is completely exposed, then the relative photographic tone may prove diagnostic (see discussion of Plate XVA in section "Relief and Tone" in Chapter 4).

On aerial photographs, the appearance of large-scale rectangular jointing is commonly associated with acid rather than basic intrusions; this may be of value in differentiating between them. For example, the rock 3, in Plate XIIB, is an acid intrusion.

If the photogeological evidence is insufficient to indicate whether a particular intrusion is acid or basic, the photogeologist should have no compunction about mapping merely an intrusion, without indicating its type.

Dykes

Because dykes are bodies of rock which have been intruded along fractures or planes of weakness in the earth's crust, many of them have some of the characteristics associated with joints or faults. Any dissimilarity between the dyke rock and the country rock may be indicated by a variation in tone, vegetation or relief.

The characteristic features of dyking are well illustrated in Plate XIIA. The country rock, which consists of Karroo sandstones and (?) mudstones, dip gently to the east-south-east. The straight negative linear feature 3–4 would normally be interpreted as a joint or fault; in fact, this feature, which continues for many miles, can be proved on evidence provided by the photographs to the north of the area illustrated, to represent a fault.

Lineament 1–2 is a straight positive feature and is collinear with 3–4. The fact that lineament 1–2 is a positive feature indicates that it has been filled with a rock more resistant to erosion than the country rock; it is therefore interpreted as a dyke. The fact that the feature 3–4 is negative does not imply that this part of the structure is not filled with dyke rock; it only implies that, if it is filled with dyke rock, the rock is less resistant to erosion than the local country rock.

Limiting ourselves solely to the evidence provided by the illustration, we are able to interpret the whole lineament 1–4 as a dyke-filled fault or joint. The structure is, in fact, a fault intruded with dolerite that apparently has a resistance to erosion intermediate between that of the (?) mudstone and sandstone, which form the local country rocks from 1–2 and 3–4 respectively.

The dykes illustrated in Plate XVIB show certain unusual features. Near points 3, 5, 6, and 9, on the stereomodel, will be found pairs of parallel relatively dark-toned lineaments forming positive features. The dark tone indicates that the rocks causing the lineaments are basic. At first glance it might be thought that in each case there were two parallel basic dykes; this is the conclusion that might be reached as a result of a single field traverse across each structure.

Further study of the stereomodel, however, produces the following observations:

(1) The four pairs of parallel lineaments are so unusual, yet so similar, in appearance, that they almost certainly have similar origins.
(2) The surface of the ground between each pair of lineaments is lower than the surface of the immediately adjacent country rock (see, for example, point 5). This indicates that the rock between the lineaments is more easily eroded than the country rock and, therefore, differs from it.
(3) Where the rock between the lineaments has been exposed by very active erosion, as at 8, it is seen to be more dark-toned, and, therefore, probably more basic, than the country rock.
(4) Where a pair of lineaments terminate, as at 1, they appear to join.
(5) Where a pair of lineaments extends into a different kind of country rock, as at 2, it produces a single wide, dark-toned lineament.

All these observations considered simultaneously lead to the interpretation: firstly, that each pair of lineaments results from one basic dyke, which is less resistant to erosion than the rock 7; and secondly, that the pairs of lineaments represent merely the peripheral edges of the basic dykes made resistant to erosion by some contact effect. (Rock 7 is a quartz-feldspar gneiss and rock 4 is a quartz-feldspathic porphyry.)

This discussion illustrates well the way in which a photogeologist arrives at his interpretation. The interpretation is not in agreement with the geological map of the area, which records the pairs of lineaments as two separate parallel diorite dykes; it is, however, in agreement with the maps of adjacent sheets, where the continuation of the structures illustrated are recorded as single dykes. Whether this proves the fallibility of photogeological interpretation or the desirability of making a photogeological interpretation before the geologist starts his field work, is up to the reader to decide.

Other dykes are illustrated in Plate XVA, point 3, and Plate VII, point 12.

EXTRUSIVE ROCKS

Basic

A basalt lava flow is illustrated in Plate XVB. The flow must have been originally continuous from points 3 to 7, but erosion of the central stream valley has cut through the basalt and exposed the underlying sediments. At points 1, 6, and 8, the non-conformable relationship between the basalt and the underlying sediments is clearly shown. Even if the area were to be covered by soil and vegetation so that the tonal contrast were lost, the landform and the non-conformable relationship would be diagnostic.

The recognition of volcanic cones is of great assistance in the interpretation of extrusive rocks. Such cones may be recognized even after considerable erosion. The landforms 1, 3, and 4 illustrated in Plate XVIA, represent extinct volcanoes in various stages of erosion. Once these volcanoes have been recognized, the explanation of the peculiar surface pattern at 2 becomes obvious; it represents a flow of lava which solidified whilst pouring down the slope in a north-west direction.

PERMEATION GNEISSES

The permeation gneiss illustrated in Plate IIB consists mainly of quartz, orthoclase feldspar, and biotite, and the dark bands differ from the lighter by virtue of the fact that they contain a higher proportion of biotite. The rock is believed to have been formed by the permeation of granitizing fluids in metasediments; some authors refer to rocks of this type as migmatites and some as banded gneisses.

Permeation gneiss similar to that illustrated is represented, in Plate XI, by rock 2. The numerous very short lineaments covering the surface of the rock outcrops such as 3, which produce a streaking or flowing effect, are caused by the mineral banding in the rock. This banding may represent the relic bedding of the original metasediments from which the rock was made; alternatively, it may result from metamorphic differentiation; in the former case, the bedding symbol is appropriate for its annotation, in the latter, the foliation symbol is more suitable (see, however, section "Photogeological Symbols", Chapter 4).

The permeation gneiss banding may still be traced on aerial photographs, even through a cover of soil and vegetation (Allum, 1961).

Permeation gneisses resemble granites in their mineral constituents and also in their general photographic appearance—compare, for example, the granite 11 with the permeation gneiss 2, illustrated in Plate XI. They differ from normal intrusive granites by virtue of the fact that because they were formed by granitization in place, they do not displace the country rock. If rock 11 (Plate XI) were a permeation gneiss, the country rock lineaments would not bifurcate near point 6 and appear to wrap around the rock, but would instead continue straight up to the rock boundary; within the rock itself, further lineaments, similar to those of the permeation gneiss area 2, would be seen, apparently collinear with those of the country rock.

In reasonably well-exposed areas, therefore, it may be possible to differentiate, on aerial photographs, between metasediments and permeation gneisses, and also between permeation gneisses and intrusive granitic rocks. The general photographic appearance and the type of lineaments present may enable permeation gneiss to be distinguished from metasediments; and the displacement or otherwise of the country rock may enable permeation gneiss to be distinguished from intrusive granitic rocks.

AUTOCHTHONOUS "GRANITES"

As the processes of granitization continue, the mineral banding of the permeation gneisses becomes progressively less pronounced until ultimately an autochthonous *granite* results. Such a *granite* may be indistinguishable from permeation gneiss on photogeological evidence. The etching effect of erosion may reveal mineralogical or textural differences, which may pass unnoticed in the field, but which may be depicted on aerial photographs as innumerable very short lineaments reminiscent of those associated with permeation gneiss.

In the field, there may be little indication in a single outcrop of the autochthonous origin of the *granite*; inspection over a larger area may be necessary to prove a gradational contact between the *granite* and permeation gneiss (Allum, 1961). Thus, the field geologist may be in danger of mistaking an autochthonous *granite* for a gneissose intrusive granite, whereas the photogeologist is in greater danger of mistaking an autochthonous *granite* for a permeation gneiss.

SUPERFICIAL COVER

The purpose of most photogeological interpretation is to assist in the understanding of the solid geology rather than the superficial cover. Unless the superficial cover makes the interpretation of the underlying rock impossible, the underlying rock itself should be mapped. Although the area south of the boundary in Plate VIB is largely covered by superficial material, the photogeological evidence is sufficient for it to be mapped as representing an area of granitic rock. Similarly, although the area depicted in Plate VIA is completely soil-covered, there is enough photogeological evidence for the areas to the east and west of the boundary to be interpreted as consisting of granitic rock—intrusive, and metasediments, respectively. In this interpretation, however, a further symbol indicating the comprehensive soil cover would be desirable.

For photogeological purposes, superficial cover is conveniently classified as residual or transported. The transported superficial cover is recognized by the following photogeological characteristics:

(1) Its almost total blanketing effect on the underlying geology; structures in the underlying geology do not normally show through transported superficial cover.
(2) The association of the cover with its means of transport.
(3) The association of the cover with diagnostic landforms such as screes, sand dunes, meander belts, river terraces, river deltas, eskers, and moraines.
(4) Its relatively sharp boundaries.

As soon as transported superficial cover is suspected, the question of its mode of transport should be considered. If, for instance, the stereomodel depicts a gently sloping pediment with flow lines extending down the slope, the superficial cover must be interpreted as being probably transported by water. If, however, the cover shows no flow lines related to slope but is associated with (say) sand dunes, there will be little doubt that the cover was transported by wind. In most cases, the photogeological evidence will be sufficient to indicate the means of transport of the cover.

Transported superficial cover is illustrated in Plate VIB, points 4 and 10; Plate VII, point 1 (extreme west); and Plate XVB, points 2 and 4.

Residual superficial cover, which includes laterites and soils, has the following photogeological characteristics:

(1) It does not tend to blanket completely the underlying geology. Because it is formed in place from the underlying rocks, its composition and texture tend to reflect in some way the underlying geology. Thus, tonal patterns or the distribution of vegetation may indicate underlying geological structures.
(2) No means of transport for the cover can be recognized.

(3) In places, it appears to have gradational boundaries with rock outcrops. This is understandable, because as it is formed from the underlying rock, there is liable to be every gradation from a thick soil cover to merely a thin layer of partly weathered rock.

Residual superficial cover is illustrated in Plates VIA, VIII, IX, and probably X. In each case, the underlying geological structures can still be interpreted.

On aerial photographs, transported superficial cover is easier to recognize than residual. If, therefore, it is desired to differentiate between them, it is advisable, firstly, to interpret the solid geology, then to delineate the areas of transported cover, and finally to study what remains, to determine whether it can justifiably be mapped as residual cover.

CHAPTER 7

Production of the Photogeological Map

ANNOTATION AND INTERPRETATION TECHNIQUES

There are several different methods of interpretation and annotation. When the utmost precision is required, the photographs may be annotated directly with water colour and a mapping pen. With this method, it is desirable to be certain of the interpretation before starting the annotations, because correction of the annotations is not easy.

When speed is important, the photographs may be annotated directly with grease pencils. This method is quick and convenient and corrections are easily made. It has the disadvantages, however, that: (a) fine detail cannot readily be traced with grease pencil; (b) the annotations are not really permanent and are liable to smudge; and (c) once an annotation covers a particular photographic image, that image cannot be seen stereoscopically.

An alternative to both the above methods is to put overlays on the alternate photographs and to annotate these with coloured pencils. The coloured pencil annotations are permanent, fairly precise, and easy to correct.

Two routines for interpretation and annotation will now be described in detail, one more suitable for unmetamorphosed and the other for regionally metamorphosed areas.

Unmetamorphosed Areas

In areas of well-exposed and clearly defined sediments where precise photogeological interpretation is possible, a satisfactory way of interpretation and annotation is as follows:

(1) Cut overlays of plastic tracing material for each alternate photograph. They should be 0·5 in. (1·25 cm) smaller than the photograph all round.
(2) Stick the overlays to the photographs with cellulose tape down their right-hand edge only. This enables the overlays to be folded to the

right, out of the line of sight, when difficulty is experienced with the interpretation and a particularly clear stereomodel is required.

(3) Write the sortie and photograph number, and your name on the margin of the overlay and mark the photograph principal point.

(4) Set up the first overlay-covered photograph for stereoscopic viewing under a mirror stereoscope with binocular attachment and transfer stereoscopically the principal points of the conjugate photographs.

(5) Trace the drainage stereoscopically to the required degree of detail. For many purposes it is reasonable to fix a lower limit of about 0·5 in. (1·25 cm) for the shortest length of drainage worth recording. If maximum accuracy is required, use high power; if time is all important, use low power.

(6) Make the photogeological interpretation of the first overlap, and record it on the overlay with coloured pencils, using the standard symbols. The stereomodel should be studied under both high and low power during interpretation but the annotations should be made under high power, if maximum accuracy is required.

(7) Stick the overlay down on left-hand edge; and remove the cellulose tape from right-hand edge. Turn the overlay-covered photograph 180°; and, using the other conjugate photograph, interpret the other half of the photograph.

(8) Complete the interpretation of all the alternate photographs in this way. Compare each interpretation with those of the surrounding photographs to ensure consistency of interpretation.

(9) Compile the interpretations on to a base map as described in Chapter 9.

For compilation purposes, it is sometimes necessary to prepare print laydowns. These laydowns, at original photograph scale, can be used for interpretation as follows:

(1) Put a mirror stereoscope on the laydown so that one photograph of the laydown can be viewed through the right eyepiece of the stereoscope.

(2) Put the conjugate photograph under the left-hand big mirror of the stereoscope.

(3) Fuse the two photographs and record the photogeological interpretation together with the topographic details directly on to the laydown in grease pencil.

If it is preferred, this method may be used with an overlay in position over the laydown as described in Chapter 9, section "No Base Maps Available", stage (8).

The advantage of interpreting directly from a laydown is that it can be used to produce a photogeological interpretation and a base map simultaneously.

It is always advisable to plan, in detail, all stages of mapping (i.e. photogeological interpretation, field mapping, and compilation) before starting work. The various methods are so interrelated that it is always possible that a particular piece of work may be used for more than one purpose.

Regionally Metamorphosed Areas

Metamorphism complicates the problems of interpretation without altering their nature. Lineaments may be observed on a particular stereopair without that stereopair providing sufficient data to indicate what they represent. Such lineaments may be of considerable length and extend across several stereopairs, any part of any one of which may provide the data necessary for their interpretation. Some lineaments may be clearly visible on print laydowns or mosaics, but be much less apparent when the relevant stereopairs are studied in detail under high power. It is therefore desirable for a geologist to be able to study each stereopair individually, then in relation to the immediately adjacent stereopairs, and finally in relation to the area as a whole. He must be able to postulate on the photographs the existence of structures and boundaries, etc., and then modify them as he obtains further data from the adjacent photographs. Finally, it is desirable that he keeps constantly in his mind the relationship between the geological structures represented on the stereopair he is studying and those of the area as a whole. For this purpose, he should be able to form a rough, uncontrolled print laydown of the photography of the whole area in a few minutes.

In order to satisfy the above requirements, a special technique is necessary. The technique described here has proved useful to the author working in metamorphosed areas in Africa.

All the edges of the photographs, except the titling strip, are trimmed off with a guillotine. By the simple process of matching edge detail, the alternate photographs are arranged into strips; they are then fastened to one another with cellulose tape. The conjugate photographs are kept separately in numerical order. The strips are then placed side by side in their correct relative position to form a rough uncontrolled print laydown; they are then numbered from north to south or east to west as the case may be. A strip near the centre of the laydown, on which the geology shows most clearly, is selected as the one on which to begin the interpretation. The other strips are piled on one another in numerical order to save space.

The photograph in the selected strip on which the geology is represented most clearly is then interpreted stereoscopically with its conjugate photographs. The best instrument for interpretation is a mirror stereoscope with parallel guidance mechanism (Plate IB); both high and low power should be used. An ordinary mirror or pocket stereoscope can be used, however.

The annotations are made directly on the photographs with grease pencils. The photographs are not removed from the strip during the interpretation; therefore, after interpreting one-half of a photograph, it is usual to turn the whole strip round in order to interpret the other half. After completion of the first photograph, the next photograph along the strip is interpreted, and so on, to the end of the strip.

The strip completed, it is then considered as a whole to see whether the interpretation is consistent from one photograph to the next. The advantage of studying each photograph on its merits, before considering it in relation to the other photographs, is that in this way attention is drawn to those areas where the interpretation is doubtful, thus reducing the danger of making an incorrect interpretation based on insufficient evidence and then transferring it from photograph to photograph.

When all the strips have been similarly interpreted, they are reinterpreted in relation to one another as follows:

(1) The side-lap between strips 1 and 2 is studied stereoscopically with a pocket stereoscope and the interpretations are compared and finally equated for both strips. Strip 1 is then studied again stereoscopically and the new information from the adjacent strip 2 is incorporated in its interpretation.

(2) The side-lap between strips 2 and 3 is next studied as in (1) above, and strip 2 is reinterpreted with the incorporation of the data from strips 1 and 3. In this way, all strips are eventually reinterpreted with the incorporation of the data from their adjacent strips, and the geological interpretation of all strips is made consistent.

When this procedure has been completed, all photographs have been interpreted twice; once individually, and once in relation to all the surrounding photographs.

THE PHOTOGEOLOGICAL MAP

In the section "Annotation and Interpretation Techniques", it was suggested that the interpretation of each photograph should be annotated either on the photograph itself or on individual overlays. Before these interpretations can be used to form a basis for further work, they must be compiled, at a convenient scale, into a photogeological map. As the photogeological map is usually required to be used at the later stages of geological mapping, it should be robust and easily amended. Perhaps the most suitable compilation is one made with coloured pencils on plastic tracing material. Care should be taken to ensure that the position and number of the principal points of each photograph are recorded, and that all roads, tracks, and other valuable data concerning the means of access are plotted. Furthermore, questions and problems arising from the photogeological interpretation should be indicated by suitable symbols and manuscript notes, so that the subsequent field work can be planned to provide the information required for their elucidation.

The choice of the scale for the preliminary photogeological map depends upon the size of the area and the maps and instruments available to the geologist. Unless instruments such as a sketchmaster or ratio projector, which enable a change of scale to be made, are available, the compilation should be made on a tracing of an uncontrolled print laydown, at photograph scale. The advantage of such a compilation is that data can be transferred to it from the photographs by direct tracing.

If, however, controlled base maps and scale-changing instruments are available and if the area to be mapped is large (e.g. a 30-minute sheet), it will be found less laborious to compile at a scale of (say) 1: 100,000. (Various methods of compilation are discussed in Chapter 9 and the geologist must select methods suitable for the instruments and base maps available.)

The photogeological map, made without the support of any field work, is liable to contain serious errors, particularly in the lithological interpretation. For many areas, however, its value is considerable, because it indicates:

(1) The main divisions of the rocks and their boundaries.
(2) Many structures that might be overlooked in the field unless attention has been drawn to them.
(3) Many structures that, because of superficial cover or some other reason, are invisible on the ground; such structures may provide the explanation of field observations that would otherwise seem incomprehensible.
(4) Special problems that only field work can solve.
(5) The means of access to the problem areas.
(6) The serial numbers of the aerial photographs required for the field work in each area.

The photogeological map enables the geologist to go into the field with a plan in his head instead of with what is euphemistically called an open mind.

CHAPTER 8

Field Mapping with Aerial Photographs

PHOTOGEOLOGY can never be divorced from field geology because, firstly, all photogeological interpretation fundamentally depends upon criteria discovered by the comparison of the stereomodel with the ground view of the corresponding area; and secondly, because the ultimate purpose of nearly all photogeological interpretations is the production of some kind of geological map, in which the data obtained from both field and photogeological work are combined.

Before starting field work, it is desirable to complete a photogeological map of the area (see Chapter 7). The exception to this general rule is if there is ample time and money for all parts of the area to be visited at least twice. Under such conditions, it might be justifiable to map the whole area, in the field, in rapid reconnaissance fashion, to obtain familiarity with the rocks and their appearance on the photographs. With this work completed, a photogeological interpretation could then be prepared prior to the main field work.

Usually, however, the time to be spent in the field has to be kept to a minimum, in which case, the order of work should be: (a) the acquisition and study of all the available geological data of the area to be mapped; (b) the production of a preliminary photogeological map incorporating the geological data from (a); (c) the field work planned in accordance with the findings of the preliminary photogeological interpretation; (d) a detailed photogeological interpretation synthesizing the field, photogeological, laboratory, and all other data of the area; and (e) the compilation of the final synthesis on to a base map.

EQUIPMENT

The list below represents the minimum photogeological equipment required by a geologist using aerial photographs in the field; it does not refer to the equipment he will probably need in his camp, such as a mirror stereoscope, grease pencils, coloured pencils, preliminary photogeological map, etc.

(1) Photographs, preferably on double weight and waterproof paper.
(2) Overlays, cut to a size slightly less than the photographs.
(3) Cellulose tape.
(4) Stereoscopic viewing board and four Bulldog clips.
(5) Pocket stereoscope.
(6) Ball-pen, press-stud type.
(7) Pencil (H).
(8) Protractor, preferably transparent.
(9) Scale.
(10) Normal geological field equipment such as hammer, compass, and notebook, etc. but excluding base map.

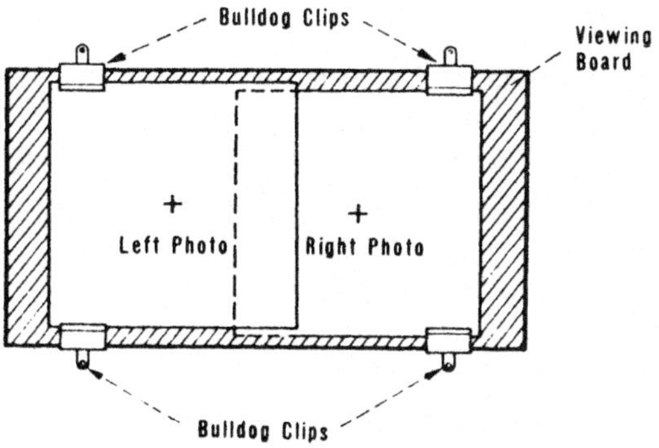

Fig. 18.

The recommended stereoscopic viewing board, for use with 9 in. × 9 in. aerial photographs, consists of a 16 in. × 10 in. rectangular piece of 4-ply wood (see Fig. 18). The Bulldog clips, in the positions indicated, hold the photographs at the most convenient separation for stereoscopic viewing with a pocket stereoscope. All the photographs required for a day's work may be kept clipped to the board, the alternate photographs on the right-hand side, in the order in which they will be required, and their conjugate pairs on the left-hand, in the corresponding order.

The use of the stereoscopic viewing board in the field saves much time, because it avoids the separation of the photographs having to be adjusted every time they are viewed stereoscopically.

PREPARING THE TRAVERSE

Each day's field work should be planned in detail in the camp during the previous evening; the photogeological map should form the basis of this planning. Decisions should be made upon: (a) the area to be investigated; (b) the best means of access; (c) the principal structural and lithological problems; (d) the outcrops that are most likely to provide the information required for the elucidation of the problems; and (e) the sortie and serial numbers of the aerial photographs required.

After the field work has been planned, the photographs required are selected and divided into alternate ones and their conjugate pairs. An overlay is attached to each alternate photograph and annotated with the principal point, the sortie and serial number of the photograph, the date of the field work, a length of main drainage, and the geologist's name.

The approximate scale of the photography is calculated, either by using the formula $S = \frac{f}{H - h}$, or by comparing the photographs with a map of known scale (see Chapters 2 and 10 respectively). The scale is recorded on the overlays by drawing straight lines representing 1 mile or 1 km.

The photographs are now studied stereoscopically to determine the most suitable outcrops to visit. (It is assumed in this discussion that the time available does not permit every outcrop to be visited.) In areas of superficial cover, the sites of probable outcrops, such as hillside gullies or incised rivers, should be indicated in pencil on the overlays.

The selection of outcrops should be such that all intruded rocks and all horizons of each group of bedded rocks are inspected at least once.

The purpose of this procedure is to ensure: firstly, that time is not wasted by visiting numerous different outcrops of the same rock body, or by making straight line compass traverses, in which it is fortuitous if outcrops are observed; and secondly, that the important but rare outcrops of largely soil-covered rock bodies are not overlooked.

After the outcrops to be visited have been selected, the best routes to be followed between the outcrops are determined, and recorded on the overlays with dotted lines.

FIELD WORK

Field Annotations

There are many advantages in recording the site of field observations on aerial photographs or overlays, rather than base maps:

(1) An observer may locate his position, and thus the position of his field observations, on aerial photographs, with much greater precision than is possible on base maps. This is because photographs provide many more topographic details than maps, and also because the stereomodel can be studied in three dimensions.

(2) As larger scale, more detailed, and more accurate maps are produced, the geological data can be transferred from the photographs or overlays to the new maps, with accuracy, without the necessity of further field work. If, however, the field observations are recorded on a base map and subsequently a more accurate base map is produced, the geological data cannot be transferred with an accuracy commensurate with that of the new map; the field work has to be done again.

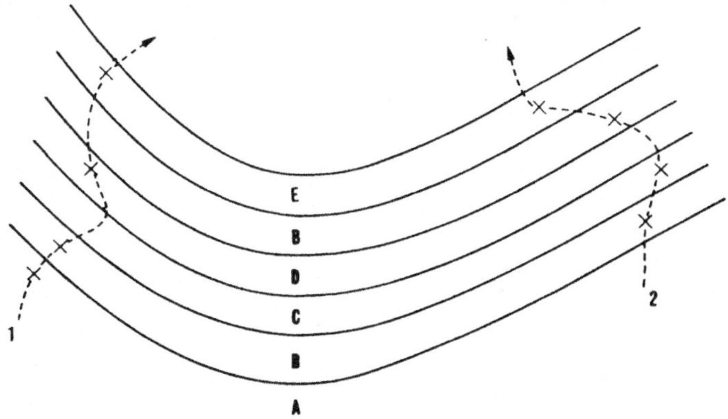

Fig. 19.

(3) If it is known precisely which photographic images correspond to particular field observations, then it may be possible to extrapolate the field observations, on the photographs, for considerable distances.

(4) As more and more field work is completed, knowledge of interpretative criteria for the particular area increases, and progressively greater value may be attached to photogeological interpretation. This does not occur, however, unless the field observations are associated in the mind of the geologist with the corresponding appearance on the stereomodel.

The importance of knowing precisely which photographic images correspond to particular field observations is illustrated by the two hypothetical traverses (1 and 2) depicted in Fig. 19.

The curved lines in Fig. 19 represent the traces on an aerial photograph of the boundaries between ill-exposed, lithological horizons. The lines 1 and

2 represent two hypothetical field traverses, and the crosses on the traverses indicate the points at which outcrops were found and specimens collected. The actual rock types are indicated by the letters *A*, *B*, *C*, *D*, and *E*; and as rocks of similar lithology may occur at different horizons in a pile of sediments or metasediments, the rock type *B* has been included twice in the illustration.

The rocks observed during the two traverses are listed in the order in which they were observed in Table 2.

TABLE 2

Traverse 1	Traverse 2
A	B
B	C
D	B
E	E

Consideration of the data in Table 2 does not suggest any correlation between the traverses; even if the traverses were plotted carefully on an accurate base map, no reliable correlation between them would be possible without additional data. The geologist would not know whether the rock *B*, seen in traverse 1, was the same horizon as the first or second rock *B* seen in traverse 2, or whether it was unrelated to either. Strike measurements would not solve the problem.

When, however, the field observations are recorded accurately on the photographs or overlays, they are automatically related to the photographic lineaments, as in Fig. 19. It is then realized: (a) that the absence of rocks *A* and *D* from traverse 2, and rocks *C* and the second *B* from traverse 1 result merely from the absence of outcrops of these rocks along the traverses; (b) that rock *B* of traverse 1 belongs to the same horizon as the first rock *B* of traverse 2; and (c) that the rocks *E* of both traverses belong to the same horizons.

The satisfactory integration of the field and photogeological data provides confirmation that the lineaments represent the boundaries of lithological horizons.

These results could not be achieved if the sites of the field observations were plotted with only approximate accuracy on the aerial photographs or overlays. It is essential, therefore, that the sites should be plotted precisely relative to the photographic lineaments.

Once the significance of the lineaments is established, they may be interpolated with great confidence between the traverses, and extrapolated with some confidence, for considerable distances, away from them.

The choice of the method of field annotations depends largely upon whether the geologist possesses his own set of photographs, or whether he has borrowed them, and has to return them eventually in an undamaged condition. There are three methods of annotation in general use:

(1) To record the sites of observations on photograph overlays, and the observations themselves in a field notebook.
(2) To record the sites of observations directly on the photographs, and the observations themselves in a field notebook.
(3) To record everything, in ink, on the photographs themselves.

Method (3) has the undoubted disadvantage that, once an annotation has been made, that particular spot can no longer be studied stereoscopically; furthermore, if errors are made and corrections are necessary, the result can be messy and difficult to understand.

Method (2) is probably the simplest and most effective, but it has the disadvantage of damaging the photographs. It is to be recommended if the geologist has his own set of photographs, which he is not obliged to return.

Method (1) is slightly more complicated to use than method (2), but it has considerable advantages. It does not damage the photographs; it permits the geologist to return the photographs to store and yet still retain a record of the precise location of his field observations; and it enables the record of the site of field observations to be stored conveniently with the corresponding field notebooks, for future reference. If the principles involved in method (1) are fully understood, there will be no difficulty in adapting them for method (2). For the rest of this chapter, therefore, unless otherwise stated, it will be assumed that method (1) is being used; this method will now be described.

An overlay is attached to the photograph, by sticking it with cellulose tape down its right-hand edge. This enables the overlay to be folded back to the right when it is required to view the underlying photograph stereoscopically without the obscuring effect of the overlay.

The principal point is marked, and a length of main drainage is traced stereoscopically on the overlay, in order to record its orientation relative to the photograph. The number of the photograph, the geologist's name, and the date of the field work are all recorded both on the overlay and in the field notebook.

The recording of the exact location of field observations is as follows:

(1) On arrival at the outcrop, the geologist should check his position by comparing the stereomodel with the ground view, first considering

the major features, and then the minor topographic details surrounding the outcrop. This is important, because many outcrops in the same environment have rather similar major features associated with them; it is the minor details, such as the number and relative positions of individual bushes near the outcrop, which enable the geologist to locate his position with precision.

(2) A line about 1 in. (2·5 cm) long, is drawn from the exact point on the outcrop at which the field observation is made, away from the route of the expected traverse. A small circle is drawn at the far end of the line, and a number, allocated to the field position, is written in the circle.

(3) The geological observations are written in the notebook, under the same number as the field location in (2) above.

The problem of cross-reference between overlay and notebook is solved by using the same note number for both. If the overlay is inspected, the date and note number indicate the correct notebook reference; if the notebook is inspected, the photograph number indicates the correct overlay, and the note number indicates the exact position on the overlay (and therefore the photograph) to which the notebook observations refer.

Geologists, using method (2), record the sites of their field observations by pricking through the photographs with a pin or needle. They write the date, and number the pin-pricks, on the backs of the photographs, using the same number as that in the corresponding notebook reference, thus providing the same cross-reference system, between notebook and photographs, as method (1) above.

It is possible to dispense with the use of a pin or needle in method (2). If a ball-pen is pressed on the sites of the field observations on the photographs, protuberances are produced on the backs of the photographs, which can be numbered like the pin-pricks. This method does not record the positions with quite the same precision as the pin-pricks, but it is considerably quicker, because the same instrument can be used both for writing in the notebook and for marking the photographs. The ball-pen is not suitable for use in wet tropical climates and, under these conditions, a hard pencil can be used instead.

Field Traverse

During field work, much time can be wasted, if a geologist loses his position on the photographs. If he knows his position to within (say) 0·25 cm or less, it is possible for him to determine quickly his precise position, by comparing the minor topographic detail on the stereomodel with the ground view. In this context, the expression "minor topographic detail" refers to such things as particular trees or bushes and their position relative to small gullies or to one another. The geologist's intention should be to establish his position with such accuracy that a pin-prick would destroy the photographic image of the very point on which he is standing.

If a geologist loses himself on the photographs, his position can usually be re-established only by walking to some topographic feature that is prominent on the ground and that can be seen on the photographs. In some areas, this can take a considerable time. It is therefore desirable to traverse in such a way that one's position on the photographs is known with considerable accuracy at all times. A method by which this can be achieved will now be described.

A starting-point for the traverse is chosen, which is fairly near the first outcrop to be visited and which is clearly defined both on the ground and the photographs; this is recorded on the relevant overlay as point 1, and a corresponding record is made in the notebook stating that point 1 represents the start of the traverse.

At the starting-point, the photographs are inspected stereoscopically and a mental note is made of the topographic details to be passed on the way to the first outcrop. The geologist then walks towards his first outcrop, stopping as and when necessary to inspect the photographs stereoscopically if he is in any doubt as to his position or if he observes any topographic features that his previous examination of the photographs had not led him to expect. For the first few days of field work with particular photography, most geologists have to inspect the photographs frequently between outcrops; as they become increasingly familiar with the scale of the photography, however, it becomes necessary to inspect it less and less.

As soon as a rectilinear feature (e.g. a straight stream-bed) that is clearly defined on the photographs and the ground is observed, its bearing should be measured; this provides a check of the orientation of the relevant photograph and overlay. A line representing magnetic north should then be drawn on the overlay.

On arrival at the first outcrop, the sites of the various observations are indicated by consecutively numbered points on the overlay, and the geological observations are recorded in the notebook under the corresponding note numbers, as described for method (1), in section "Field Work—*Field Annotations*".

The observations are made and specimens collected in the same manner as for conventional field mapping. When specimens are collected, it is recommended that the fact is indicated by a special symbol written against the relevant numbered point on the overlay; this enables the relevant notebook reference to be found quickly; the description and number of the specimen is, of course, included in the notebook record.

Before leaving each outcrop, the photographs should be inspected stereoscopically to determine, if possible, the regional significance of the field observations, and to decide whether a change in the planned field work is required. The traverse then continues from outcrop to outcrop in this manner.

If the distance between outcrops is considerable, care should be taken to inspect the photographs stereoscopically at regular and frequent intervals to provide an accurate determination of position; this position should be recorded on the overlay with a number, even though no geological observation is made; the notebook comment should then read "position only".

If a geologist checks his position on the photographs frequently, the time spent on each check is very short because the position is known to within a short distance and only a minute portion of the stereomodel has to be inspected. Less frequent determinations of position usually result, in the end, in a greater total time spent on the operation. Furthermore, if a geologist loses his position on the photographs, he may not find it again, so that much of the day's work is wasted.

Certain modifications of this method of field work may prove necessary when the photography is more than about 7 years old and when the outcrops are separated by considerable areas in which there are few topographic details distinguishable on the ground and the photographs.

If the photography is old, the vegetation may be a hindrance rather than a help in the determination of position. In the interval since the photography was taken, old trees and bushes may have died, and new ones grown. With this kind of photography, it is necessary to ignore the vegetation and to concentrate on relief for the purpose of locating position.

If there is a considerable area between two outcrops, in which there are few topographic details, the bearing and distance between the outcrops should be measured on the photograph, and a compass traverse made from the one to the other. This has proved satisfactory in areas of moderate relief but it is possible only if the scale and orientation of the photographs have been ascertained and recorded as recommended.

If the outcrops to be visited are very far apart or if the area is one of considerable relief in which the bearings are grossly unreliable, the traverse should be divided into several short lengths between topographic features that are visible on the photographs and the ground and that lie on the approximate line of traverse. The bearing and distance between these topographic features are found from the photograph, as before. An added advantage of this system is that it enables a route to be chosen that is convenient for walking.

In some areas, outcrops are difficult to find in the field, although their photographic images are distinct. Such conditions do not necessitate a radical change in the method of field work; all that is required is some means of finding the outcrop in the field. The solution is to choose, on the stereomodel, some topographic feature that is very near to the required outcrop and that is likely to be easily recognizable on the ground, and to traverse to this, by bearing and compass method. When the feature is found, the outcrop will be found, and the field work can continue as described.

On returning to the camp after the day's field work, the photographs in the immediate vicinity of the field work area should be restudied; the aim should be to amend the photogeological interpretation in the light of the field observations. All geological structures observed in the field should be extrapolated on the aerial photographs as far as possible, and the photogeological map should be amended with the new interpretation and the field observations. The rock names used for the photogeological map (see section "Generalized Photogeological Legend", in Chapter 6) should be replaced by rock names based on field observations and inspection of the hand specimens.

Henceforth, the amended photogeological map becomes, and will be referred to as, the **working map.**

The resulting working map should then be used to plan the following day's field work, as described in the section "Preparing the Traverse".

FINAL SYNTHESIS OF GEOLOGICAL DATA

After the completion of the field work, more geological data are obtained from laboratory studies of the specimens collected. The names given to the rocks in the field are amended in accordance with the results of the study of thin sections. A final synthesis of all the geological data of the area is then made. This synthesis is carried out by making a detailed photogeological study under a mirror stereoscope. There are several reasons for this:

(1) The acquisition of the field and laboratory data, although solving some problems, will have raised others, which photogeological interpretation may elucidate.

(2) The field and laboratory data will have led to the discovery of structures not included on the preliminary photogeological map; these will have to be interpolated or extrapolated as far as possible on the aerial photographs, and their regional significance considered.

(3) Study of the stereomodel assists in the synthesis of the geological data, because it makes the field notes more meaningful, and keeps the relationships between the various rock groups constantly in the mind of the geologist.

When this final detailed photogeological interpretation is completed, the working map is amended again in accordance with it.

CHAPTER 9

Compilation

AFTER completing the field, photogeological, and laboratory studies, the geologist faces the problem of transferring his synthesis of all the geological data, which he has recorded on his working map (see Chapter 8), to a base map at a predetermined scale.

The different circumstances that may face a geologist working in the underdeveloped countries of the world are as follows:

(1) There is a reliable base map.
(2) There is an unreliable base map, which, however, contains certain topographic features such as main rivers or roads that have been accurately surveyed.
(3) There are no suitable base maps, but there is a regional map of small scale (say 1: 500,000).

Each of these different conditions poses different problems to the geologist; they will be discussed separately.

The methods described may be used in many different combinations. It is up to the geologist to select from them the combination most suitable for his circumstances.

All map sheets, however they are compiled, should contain the following:

(1) Name and number.
(2) Two lines of both latitude and longitude, even if only approximate.
(3) Scale, even if only approximate. This should be in figures as a ratio and also as a line representing (say) 5 miles or 10 km.
(4) If the map is not completely controlled, notes should be included at the bottom of each sheet stating that it is uncontrolled, that it should not be used for the measurement of bearings or distances, and giving the date, outline method of compilation, and name of compiler.

All maps compiled from aerial photographs should also include the principal point positions and the sortie and serial numbers of the photographs.

RELIABLE BASE MAPS AVAILABLE

The technique of producing maps from aerial photographs is a branch of photogrammetry; the development of photogrammetric methods has, more than any other factor, made the production of controlled, accurate, large-scale maps quicker and less costly. Every year additional areas are mapped by these methods; it becomes ever more probable, even in the least developed countries, that controlled base maps will be available for geological compilation.

When maps are produced from aerial photographs, it is possible to trace topographic features, and in particular the drainage, in detail, and with great accuracy of shape. Such maps are ideal for geological compilation, because the problem of plotting geological data on to a map is largely a matter of relating the topographic detail on the base map with that in the field, or on the corresponding photograph or photograph overlay.

In transferring geological data from a working map, annotated photograph, or overlay to a base map, much time can be saved by the use of a sketchmaster or a ratio projector.

A ratio projector consists of an illuminated movable plate, a movable lens, and a sheet of glass at convenient drawing height. The base map is placed on the glass; the working map is put on the illuminated plate; and the image of the working map is projected through the lens, glass, and base map. The instrument can be used for either enlargement or reduction, and the scale of projection is variable over a fixed range. The maximum scale change depends upon the design of the instrument, but for the Grant, is about 5:1.

The method of use is to adjust the scale of projection until the topographic details of the working map, etc. coincide with those of the base map. If the shapes of the topographic details on the working map and base map are correct, a satisfactory match of topographic detail can be obtained, providing sufficiently small areas are matched at any one time. These conditions are likely to apply, at least so far as drainage is concerned, if both the working map and base map have been prepared from aerial photographs. Once the topographic details are matched, the geological annotations, within the matched area, are traced with coloured pencils on to the base map.

As the scale adjustment is very quick with the ratio projector, the method provides a rapid and convenient way of transferring geological annotations from a working map, an overlay, or an annotated photograph, to a base map. It does, however, suffer from the disadvantage that, unless the base map is very translucent and the original annotations on the working map are very clear, the projected image seen through the base map may be obscure and difficult to trace. It is therefore recommended that, before using this method, the base map should either be photographed on translucent bromide paper or traced on a plastic sheet.

A sketchmaster (Plate IIIA) is a portable instrument, which works on the camera lucida principle. It consists of a plate, a large mirror, and a half-silvered mirror. The instrument stands on the base map at which the user looks via the half-silvered mirror. The working map is put on the plate and its image is transmitted to the user by reflection off, first the large mirror, and then the half-silvered mirror. Thus, the user sees an image of the working map superimposed on the base map. Scale variation is achieved by altering the length of the legs, which alters the apparent size of the base map without affecting that of the working map. Depending on the length of the legs, different lenses have to be inserted between the half-silvered mirror and the base map.

The method of use is to alter the lens and the length of the sketchmaster's legs until the topographic detail of the base map matches the topographic detail of the virtual image of the working map; the geological annotations are then plotted on the base map in coloured pencil by drawing on the virtual images of the working map annotations.

This method works very well with both opaque and translucent base maps; it is slightly slower than the ratio projector method, because the sketchmaster takes longer to adjust for scale change. The maximum scale change possible with the sketchmaster, illustrated in Plate IIIA, is about 2:1.

The sketchmaster is also of value for the purpose of transferring geological annotations from aerial photographs or their corresponding overlays at one scale to either translucent or opaque base maps at a smaller scale. It is therefore useful for the compilation of the preliminary photogeological map.

The robustness and portability of the sketchmaster makes it a useful instrument in all geological offices where aerial photographs are used.

If a sketchmaster or ratio projector is not available, it is possible to transfer the geological annotations from the working map to the base map merely by comparison of topographic detail. In practice, the accuracy with which this can be done depends upon the amount and distribution of the topographic detail, and in particular the drainage, on the base map.

Photographic Compilation

An entirely different method of compilation, based on photographic techniques, has been used in Australia by the Bureau of Mineral Resources (Anon., 1963). It should be of value when very large areas are to be compiled at a small scale. An outline of the method is as follows:

(1) The photogeological interpretation, the drainage, and other topographical details are annotated on individual overlays of alternate photographs.
(2) Control points are transferred, with a pantograph, from the base map to the compilation sheet, at the required reduced scale (say 1 : 250,000).
(3) The individual overlays are photographed with a 35 mm camera.
(4) Each 35 mm negative is enlarged until its control points have the same spacing as the corresponding points on the compilation (2). The negative and compilation are then at the same scale. A print is made at the new scale on Kodak Projection Thin or Ilford 23P paper.
(5) The prints are put on the compilation sheet with their control points agreeing with the corresponding points on the sheet, and are stuck down in that position.
(6) The compilation is checked, fair drawn, and if necessary hand coloured.

UNRELIABLE BASE MAPS AVAILABLE

For some areas, the only base maps available are those made before aerial photographs existed. It may be found, with some of these maps, that the main drainage and any road or rail system have been accurately surveyed but that the minor drainage, which is so vital for the plotting of geological annotations, is unreliable. To ascertain which parts of the map have been accurately surveyed, compare the topographic details of the map and photograph; the "shapes" of the topographic details, such as roads, tracks, railways, and main drainage, on the map, will agree with those on the relevant aerial photograph, only on those parts of the map that have been accurately surveyed.

The geologist may use the surveyed parts of such maps to provide partial control for the production of his own semi-controlled base map at approximate photograph scale. This base map may be conveniently used for the production of the preliminary photogeological map and, if a ratio projector is available, it may be reduced in scale and used for the final compilation also.

A method for producing the semi-controlled base map is as follows:

(1) Annotate the alternate photographs with grease pencils with all the topographic data required on the map.
(2) Make the alternate photographs into a print laydown as described in Chapter 1.
(3) Compare the "shapes" of the main drainage, roads and tracks, etc. on the print laydown with those on the map, and determine which areas have been accurately surveyed, as described above.

(4) Select points, well distributed over the map, that are clearly defined both on the map and on the print laydown and that occur on the surveyed features of the map as determined by (3) above. Note the latitude and longitude of these points with a grease pencil on the print laydown.

(5) Select a straight length of one of the surveyed features on the map and, with a protractor, measure its bearing.

(6) Using the known bearing of the straight feature (5), draw a straight line with a grease pencil on the print laydown to represent the north direction. This direction will be only approximately true on the laydown.

(7) Using the direction for north and thus east obtained in (6), interpolate between the points of known latitude and longitude obtained in (4), to find points which lie on the graticule of the map; i.e. if the graticule of the map to be is at a spacing of 5 minutes, then interpolate to find points which lie on the 05, 10, 15, etc. minute lines. Mark these points on the print laydown with a grease pencil.

(8) Draw lines with a grease pencil joining the points obtained in (7) to form a graticule on the laydown. Because of scale variation over the photographs, this graticule will appear imperfect; each graticule line will consist of a number of short, straight, but not quite parallel lines.

(9) Put sheets of plastic tracing material over the print laydown and trace all annotations. The resulting map is a semi-controlled compilation at approximate photograph scale. Bearings and distances are only approximately accurate, but the shape of the minor topographic details, particularly the drainage, will be very good. It has the further advantage that geological annotations may be transferred to it from aerial photographs or their overlays by direct tracing. It forms a very satisfactory base for the preliminary photogeological map, unless the area to be mapped is extremely large or mountainous.

If the area to be mapped is extremely large, or if it is required to produce a base map for the final compilation, then stage (9) above should be ignored, and the compilation should be made at a smaller scale as described below.

(9a) Draw a graticule on plastic tracing material at the scale required for the final map and at the same interval as on the print laydown.

(10) Transfer all topographic data including the principal point positions from the laydown to the compilation sheet (9a), using the graticule as approximate control.

If a ratio projector is available, a recommended method is to divide the print laydown into 5-minute "squares" and project each of them in turn on to the corresponding square of the compilation sheet. The topographic detail can then be traced on to the sheet. If the scale of the compilation is considerably smaller than that of the laydown, the imperfections of the laydown graticule (8) will be unimportant, and an almost perfect match of the 5-minute "squares" will be obtained.

The map which results from this process will have very detailed drainage, which is accurate in shape, and which thus appears very convincing; it should always be remembered, however, that the map is only semi-controlled and that distances measured on it will not be quite accurate.

NO BASE MAPS AVAILABLE

In areas for which no base maps are available, the geologist must either make his own base map by photogrammetric methods, or else use entirely uncontrolled compilations, such as print laydowns.

The photogrammetric method by which the geologist can make his own map is somewhat laborious, and is only to be recommended when planimetric accuracy is very desirable, and when the area is not very large. The method is described in Chapter 10.

If the geologist is responsible for a large area of regional mapping, such as a 30-minute sheet, he would be well advised to leave the production of controlled maps to the surveyors and to make use instead of some form of uncontrolled photographic compilation. His efforts should be concentrated on relating his geological data as accurately as possible to the detailed drainage.

One method which has proved both quick and satisfactory is as follows:

(1) Make an uncontrolled print laydown with the alternate photographs, as described in Chapter 1. (It is desirable to lay down the photographs of the whole area in one piece, because it is not easy to fit uncontrolled compilations to one another.)

(2) Using whatever data are available from small-scale maps, draw the boundaries of the area to be mapped with a grease pencil on the laydown, and note the approximate latitude and longitude of any points which are recognizable both on the laydown and the small-scale map.

(3) By interpolation between the boundaries and any points of known latitude and longitude from (2) above, draw an approximate graticule in grease pencil on the laydown.

(4) Cut sheets of plastic tracing material to a size suitable for map sheets. (It is desirable that the maps should fit easily into any map drawers which are available.)

(5) With a straight edge and grease pencil (not the same colour as the graticule) draw rectangles on the laydown of a size suitable for plotting on the map sheets (4).

(6) Remove from the laydown the first part of the area to be plotted, and put it under a map sheet.

(7) Trace on the map sheet, in ink or pencil, the area outline, and record and number the photograph principal points positions.

(8) Place a mirror stereoscope on the map sheet, in a convenient position for viewing, through the right eyepiece, one of the photographs of the underlying print laydown. Place the conjugate photograph, on top of the map sheet, in the position for viewing through the left eyepiece of the stereoscope. Fuse the two photographs, and annotate the topographic detail on to the map sheet in pencil in exactly the same way as when a single stereopair is studied with an overlay on the right-hand photograph. After completing the whole map sheet in this manner reattach the portion of the laydown to the main laydown.

(9) Complete all the map sheets in this manner.

(10) Ink in the topographic annotations. Do not draw a full graticule, but draw short lines perpendicular to the boundary lines, to indicate the position of the approximate graticule.

It will be noticed that this method of drawing the topographic detail stereoscopically on to the map sheet could have been used with the previous compilation method, see section "Unreliable Base Maps Available", stages (1)–(10). The choice of method depends upon the particular circumstances, such as the ease with which the topographic detail can be interpreted through the tracing material and the time available.

It would be possible, with the above method (1)–(10) in this section, to interpret and record the photogeological interpretation simultaneously with the production of the uncontrolled base map. This, however, is only to be recommended if the interpretation is simple (see Chapter 7, "Annotation and Interpretation Techniques").

For rapid reconnaissance photogeological interpretation, in which all other considerations are sacrificed to speed, use may be made of published, uncontrolled print laydowns for compilation purposes.

The majority of the print laydowns produced by the Directorate of Overseas Surveys consist of 30-minute sheets at a scale of about 1: 120,000. On most of these laydowns, the main drainage and roads show clearly and the photograph principal points can be discerned.

The quickest way to use such print laydowns for compilation purposes is as follows:

(1) Annotate the principal points and the topographic details, such as main roads and main drainage, non-stereoscopically on the laydown, with a grease pencil.
(2) By simple comparison of principal point positions and topographic details, transfer the annotations from the working map, aerial photograph, or aerial photograph overlay, to the laydown. The annotations are made with grease pencils.
(3) Put a sheet of plastic tracing material over the laydown and trace in coloured pencil the laydown boundaries, photograph principal point positions, topographic details, and geological annotations.

This type of compilation contains many errors, such as: those scale errors inherent in aerial photographs; those resulting from the making of an uncontrolled print laydown; those resulting from the annotations of drainage, etc., non-stereoscopically; and those resulting from the transference of geological annotations by visual comparison of topographic detail.

Nevertheless the great speed of this method, by which geological annotations at photograph scale may be compiled at a much smaller scale, makes it valuable for certain reconnaissance purposes.

CHAPTER 10

Photogrammetry for Geologists

A GEOLOGIST may produce a controlled planimetric map, using only his usual drawing instruments and the aerial photographs. The method is rather laborious, and is unsuitable for areas covered by more than (say) four strips, each of ten photographs; but it is of value to geologists responsible for mapping in detail areas for which no base maps are available.

The various stages of the work are: (a) the production of a minor control plot for each strip of photography; (b) the reduction of each minor control plot to the same scale; (c) the addition of subsidiary control; and (d) the plotting of the topographic detail.

The methods described are susceptible to considerable elaborations and refinements (see Admiralty, 1948; American Society of Photogrammetry, 1966; Kilford, 1963; and Trorey, 1950). The whole process depends upon the assumption that the angle subtended by two points on an aerial photograph at the principal point is very nearly correct; i.e. that it is virtually the same as the angle that would be measured by a theodolite set up on the ground at the point corresponding to the principal point of the photograph, and sighting first one point and then the other.

The reader is recommended to read Chapter 2 again, in which the geometry of aerial photographs is discussed, before attempting to make a minor control plot.

MINOR CONTROL PLOT

In the following work, all points must be transferred and all lines drawn with the greatest possible accuracy. Errors do not tend to cancel out but rather to build up; thus, gross errors in the early stages soon make further progress impossible.

It is assumed that the area to be mapped is not larger than that covered by four to five strips of seven to eight photographs each. If larger areas are to be mapped, further refinements of technique are necessary, such as those described for the transference of principle points, and internal adjustment, by the Admiralty (1948) and Kilford (1963).

(1) Decide upon the boundaries of the area to be mapped and try to include at least two points (*A* and *B*), of known position on the ground, that are recognizable on the aerial photographs. Work will be saved if these two points fall within a single strip of photography; they must, however, be a considerable distance apart, if they are to provide orientation and scale control for the compilation; ideally, they should be at opposite ends of a single strip of photography.

With a needle, prick through points *A* and *B* on all photographs on which they occur; annotate the prick-marks with the relevant letters in grease pencil.

(2) Select all the photographs required for the area, and form them into a rough print laydown, without sticking them down. Check that the overlaps are more than 50 per cent and that there are no gaps in the photography or other fundamental errors (see Chapter 1).

(3) For each strip of photography, cut a piece of plastic tracing material rather larger than the strip. Place each piece over its corresponding strip, and lightly trace the positions of the four corners of the strip.

(4) Cut another piece of plastic tracing material sufficiently large to cover the whole area of the laydown (stage (2)) and still leave a generous margin all round. This will be known as the compilation sheet.

If points *A* and *B* (see stage (1)) are on the same strip of photography, their distance apart should be calculated at a convenient scale for the compilation, which should be close to the mean scale of the photographs. They should then be plotted on the compilation sheet at this separation, in a position such as to allow the whole compilation to be drawn conveniently on the sheet. If, however, points *A* and *B* are on different strips, then point *A* only should be plotted at a convenient position on the compilation sheet.

(5) Select tie points (*T.P.*). The purpose of these is to enable adjacent strips to be brought to the same scale and plotted as a single compilation.

They should be minute, clearly defined points of detail that occur on both strips and that form satisfactory intersections with the principal points of the three nearest photographs of each strip.

All the internal strips are adjacent to two other strips and thus require a minimum of four tie points, one near each corner of the strip. It is desirable, however, to choose tie points on every alternate overlap.

(6) After selecting a tie point on one photograph, prick it through with a needle, and transfer it stereoscopically under high power to the adjacent photographs of the same strip, as described in section

"Transference of Principal Points", Chapter 3. Find the equivalent point on the adjacent strip; prick it through; and transfer it stereoscopically to adjacent photographs of the same strip. If the prick-marks are rubbed over with yellow grease pencil and then wiped, some grease will be left in the holes, making them very distinct. (It is recommended that all lines be drawn with a needle and straight edge and that all tie points, transferred principal points, minor control points, base lines, and radial lines, marked on the aerial photographs for the construction described below, be emphasized with coloured grease pencil in this manner.) Draw small circles round the tie points with grease pencil.

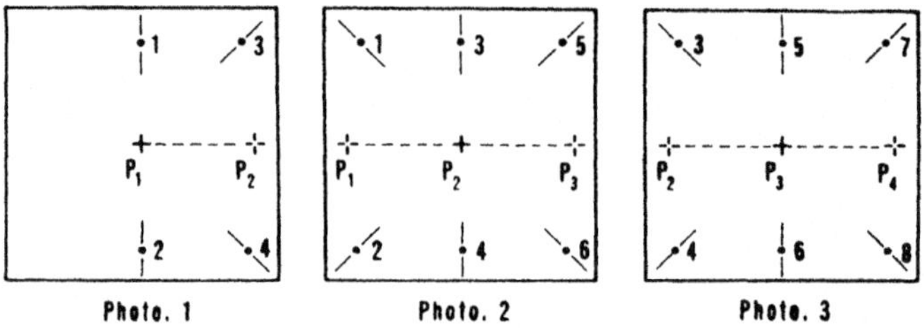

FIG. 20.

(7) Prick through the ground control points, such as *A* and *B*, and draw a small circle around them.

(8) Transfer the principal point from each photograph to its adjacent photographs, stereoscopically (see Chapter 3). This must be done very carefully.

(9) Base line each photograph by joining its principal point to the two transferred principal points (see Fig. 20).

(10) Select the minor control points. These should be minute points of detail, similar to the tie points (see stage (5)); the tie points already chosen should be used also as minor control points, if convenient. In order to obtain the best intersections, the distance of the minor control point from the nearest principal point should be about the same as the length of the principal point base line—e.g. in Fig. 20, the distance 5–P_3 is about the same as P_2–P_3.

It is essential that every point, except those such as 1 and 2 on the terminal photographs of the strip, should appear on three successive photographs; thus in Fig. 20, minor control points 3 and 4 appear on photographs 1, 2 and 3; similarly, points 5 and 6 appear on photographs 2, 3, and 4. It will be seen from Fig. 20 that all the photographs, except those at the two ends of the strip, have six minor control points.

The procedure is that minor control points 1 and 2 on photograph 1 are selected, pricked through, and then transferred stereoscopically to photograph 2; then points 3 and 4 on photograph 2 are selected, pricked through, and transferred stereoscopically to photographs 1 and 3, and so on to the end of the strip.

(11) Draw rays from the principal point to the minor control points on each photograph, as shown in Fig. 20. They need extend only a short distance on either side of the points.

Draw similar rays to any tie points or fixed points such as *A* and *B* (see stage (1)).

(12) Place the plastic tracing material for the first strip (see stage (3)), henceforth called the minor control plot, on a table. Slide the first photograph under it, into the position indicated by the traced corners (see stage (3)). Place weights on top of the plot and photograph. Trace the positions of the principal points P_1 and P_2, and join them with a straight line. (All drawing on the minor control plots or compilation sheet is done with a sharp 3*H* or harder pencil.) It is advisable to continue the line P_1–P_2 somewhat beyond P_2 as shown in Fig. 21.

Trace the rays from P_1 through the points 1, 2, 3, and 4 and any tie points and fixed points.

(13) Remove photograph 1, and put photograph 2 under the plot, with the photograph base line P_1–P_2 lying along P_1–P_2 on the plot, and the point P_2 on photograph coincident with corresponding point on the plot. (In general, P_1 on photograph will not now be exactly coincident with P_1 on the plot. This is because the scales of the photographs 1 and 2 will not normally be identical. The scale of the whole plot is determined by the length P_1–P_2 on the first photograph.) Fix photograph 2 in position with weights.

Trace the base line P_2–P_3 and extend it beyond P_3. Do not trace the position of P_3.

Trace rays from P_2 through the minor control points 1, 2, 3, 4, 5, and 6 and any tie points and fixed points. The rays to points 1, 2, 3, 4, *T.P.* and *A* will intersect those already traced from photograph 1; these intersections represent the relative positions of the points at the scale of the plot. (Thus *A′* represents the position of *A*, relative to the other points, at the scale of the plot.)

(14) Remove photograph 2, and put photograph 3 under the plot with base line P_2–P_3 on photograph coincident with the line P_2–P_3 on the plot. Keeping these lines superimposed, slide the plot over the photograph until the ray on the photograph from P_3 through point 3 passes through the intersected position of this control point on the plot (see stage (13)). If everything has been done correctly, the ray on the

photograph from P_3 through point 4 should pass through the intersected position of point 4 on the plot; if it does not, there is probably a mistake in the previous work and this should be checked. If the check shows that everything has been done accurately, the plot should be moved so that the rays pass as closely as possible to points 3 and 4 (i.e. the best mean fit), whilst still keeping the base lines coincident. Fix the photograph with weights. Trace the position of P_3, and trace the rays from P_3 to the minor control points 3, 4, 5, 6, 7, and 8 (see Fig. 21).

The rays to points 5 and 6 from photographs 2 and 3 will intersect; these intersections represent the plan positions of the points.

Trace the line P_3–P_4 and extend it beyond P_4, but do not trace the position of P_4.

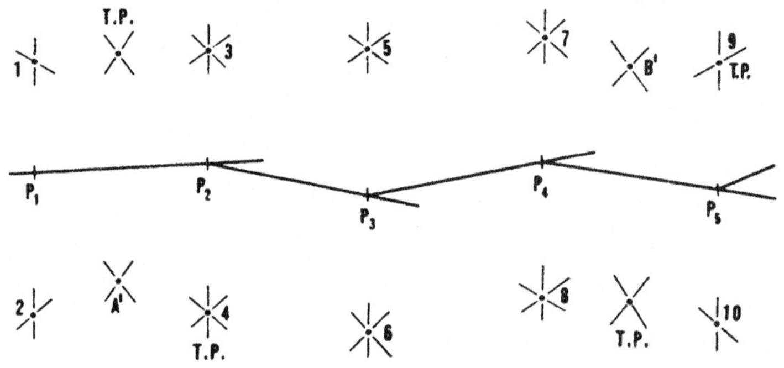

FIG. 21.

(15) Remove photograph 3, and put photograph 4 under the plot; continue as for stage (14), photograph by photograph, until the whole strip has been completed. The completed minor control plot will be similar to that shown in Fig. 21.

Points A' and B' are the fixed points A and B (see stage (1)) as they appear on the minor control plot. They may be trigonometrical survey points or merely points of known position, taken from a map.

(16) Complete a minor control plot for each strip of photography.

The scale of each minor control plot is determined by that of its first photograph; thus, in general, the scale of each will be slightly different. It is therefore now necessary to bring all minor control plots to the same scale.

If both points A' and B' (equivalent to points A and B, see stage (1)) are on one minor control plot, the problem resolves itself into that of bringing this plot to the required final scale, and then bringing all the other plots to the same scale. (If points *A'* and *B'* are on different minor control plots, turn to stage (21).)

A method will now be described for bringing the minor control plot, which contains both points *A'* and *B'*, to the required final scale.

(17) Put a piece of plastic tracing material over the compilation sheet (see stage (4)), and prick through points *A* and *B*. Join points *A* and *B* with a pencilled straight line, and extend the line beyond *B*. Put the tracing material over the minor control plot, with point *A* coincident with point *A'* on the plot, and line *AB* lying along *A'B'*. Prick through *B'*. With centre *B'* and radius *BB'*, describe a circle. From *A* and *A'* draw a tangent to this circle that touches it at *C*; join *B'C*. The diagram is now as shown in Fig. 22. (*B'* may be on the other side of *B* but that does not affect the description or use of the diagram.)

Attach the diagram to a drawing board and place the minor control plot to be corrected on top of the diagram with point *A'* on the plot coincident with point *A* and *A'* on the diagram. Put a pin through point *A'*, so that the plot is free to rotate about point *A* and *A'* on the diagram.

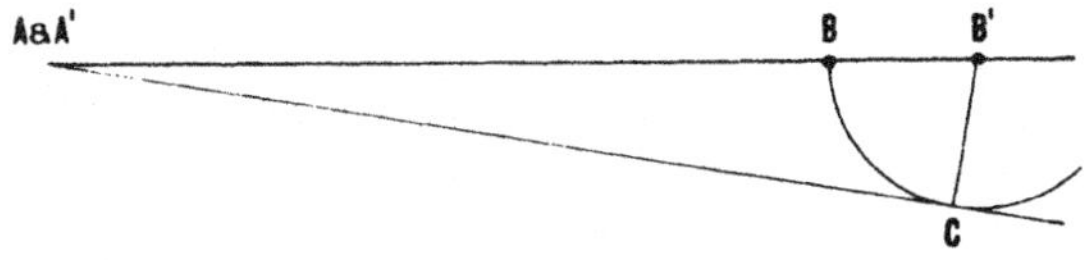

FIG. 22.

(18) To correct the position of any point on the plot, rotate the plot about point *A* and *A'* until the line joining *A* and *A'* to *B'* passes through the point. With dividers or compasses, measure the shortest distance from the point to be corrected to the tangent line, and mark off this distance from the point, in the direct line towards *A* and *A'*; this is the new required position of the point. (If the scale of the plot were being increased instead of reduced, point *B* would fall on the other side of *B'* in Fig. 22, and the correction would be made in the direct line away from *A* and *A'*.) Thus, for the point *B'*, the shortest distance to the tangent is *B'C*; this distance, measured from *B'* towards *A* and *A'*, will bring *B'* into coincidence with *B*, as is required.

Correct the positions of all the points in the plot in the manner described in stages (17) and (18). (Some points may be further away from *A* and *A'* than B'; in this case, it will be necessary to extend the tangent beyond *C*.)

(19) The method of scale change described in stages (17) and (18) may be applied whenever two constructions at different scales have two points in common.

It may be used to bring an uncorrected plot to the same scale as an adjacent corrected plot. For this purpose, the common, end tie points of the corrected and uncorrected plots take the place of A and B and A' and B' respectively, and a new diagram, similar to Fig. 22, is drawn.

Bring all plots consecutively to the correct scale in this manner.

(20) Place the compilation sheet over the minor control plot that contains the now corrected points A' and B', with point A coincident with A' and B coincident with B'. Trace the corrected positions of all principal points, tie points, and minor control points. Then place the compilation sheet over an adjacent minor control plot, with the end tie points traced off the first plot in coincidence with those of the second minor control plot, and trace the corrected positions of the principal points, and minor control points of this plot. Continue this process, plot by plot, until all the corrected points have been traced on to the compilation sheet, at the final required scale.

This completes the minor control plot and the next section of work consists of producing subsidiary control points—see below.

(21) If points A' and B' are on different minor control plots (see end of stage (16)), then it is necessary to proceed as follows.

Bring all the minor control plots to the scale of any one of them (see stage (19)).

Trace all plots on to the compilation sheet. The method is as stage (20), except that, as there is no point B on the compilation sheet (see stage (4)), the orientation of the tracing is unknown.

(22) Calculate the distance AB (see stage (1)) at the final compilation scale, and measure this distance from point A, along AB', on the compilation. Prick through this true required position of point B.

(23) Bring the scale of the whole compilation to that determined by the length AB (see stage (19)).

SUBSIDIARY CONTROL POINTS

The reader is recommended not to decide upon the amount of subsidiary control necessary until he has read the section "Detail Plotting" below.

The method of producing subsidiary control points is as follows:

(1) On each stereopair, select clearly defined points of topographic detail on the common overlap. Prick through these points on both

photographs. (In practice, these points will be selected and pricked through on one photograph, and then transferred stereoscopically under high power to the other (see "Transference of Principal Points", Chapter 3).) The desired spacing of the subsidiary control depends upon the topography; the more rugged the country, the closer should be the spacing.

On each photograph, draw rays from the principal point to the subsidiary control points.

(2) Select a photograph, and slide it under the compilation until its principal point coincides with the corresponding point on the compilation and its base line lies along the line joining the two corresponding principal points on the compilation. Trace the control point rays on to the compilation.

(3) Repeat stage (2) with the conjugate photograph. The intersections of the rays then represent the planimetric position of the subsidiary control points.

(4) Repeat stages (2) and (3), photograph by photograph, until the planimetric positions of all the subsidiary control points have been obtained.

DETAIL PLOTTING

If there is sufficient subsidiary control and if the scale of the compilation approximates to that of the photography, topographic detail may be transferred from each photograph to the compilation by direct tracing. The method is to place the photograph under the compilation, in the position of the best fit between three adjacent control points on the photograph and the corresponding points on the compilation; the topographic details within the triangular area enclosed by imaginary lines joining the control points are then traced; the position of the compilation on the photograph is then adjusted for the next triangular area, and the process is repeated.

If some of the topographic details on the photographs are not distinct, it may prove advantageous to annotate them before doing the tracing.

If there is a considerable scale difference between the compilation and the photographs, the use of plotting instruments such as a ratio projector or sketchmaster (see Chapter 9 and Plate IIIA) is desirable.

By using one or other of these instruments suitably adjusted, it may be possible to obtain the image of the principal point, the transferred principal point, and two minor control points of a photograph (e.g. points P_1, P_2, 1 and 3 of photo 2, Fig. 20) in coincidence with the corresponding points on the compilation; under these circumstances, where the ground is fairly flat, the topographic details on the photograph may be traced straight on to the compilation without the production of subsidiary control. For very hilly country, which produces considerable scale change, this will not be possible, and subsidiary control will be necessary.

If there is a large amount of topographic detail to be accurately plotted, a radial line plotter can be of great assistance. This instrument consists of a mirror stereoscope, a drawing attachment, and two plastic arms that extend radially from the principal point of each photograph. When the observer looks through the stereoscope, he sees, superimposed on the stereomodel, the two plastic arms intersecting each other. There is a mechanical linkage between the radial arms and the drawing attachment, such that, when the instrument is correctly set up, the drawing instrument is in the plan position represented by the intersection of the plastic arms.

Plotting with this instrument consists of moving the drawing attachment in such a way that the point of intersection of the plastic arms moves along the detail to be plotted. The plastic arms take the place of rays drawn from the principal points; the result is as though every point of detail had been intersected from the principal points, as described in the previous section.

This is the most sophisticated and satisfactory instrument for plotting topographic detail that the geologist is likely to use. It can, of course, be used for plotting photogeological interpretations or any other interpretations with which the photographs have been annotated.

MEASUREMENT OF HEIGHT DIFFERENCES

It was shown in Chapter 2 that: (a) the difference in height between two points on a stereopair is related to the difference in parallax between the points; and (b) the difference in parallax between the two points is equal to the difference in the distance separating the corresponding images of the points on the two photographs, when they are correctly set up for stereoscopic viewing (i.e. in Fig. 8, the parallax difference of the points b and t is given by the difference between the length d_1 and d_2). Thus, if we had an accurate method of measuring d_1 and d_2 and a formula linking differences in parallax with differences in height, we would have a method of calculating differences in height. The rest of this chapter is mainly concerned with these problems and their solution.

If the photographs are tilted, errors in the measurement of height differences can arise; these errors are liable to be serious, if the points to be compared are widely separated, particularly if the direction of separation is perpendicular to the base line. In general, it is advisable to determine heights by interpolation between points of known height. For a description of height correction techniques, Kilford (1963) should be consulted.

The measurement of height differences from aerial photographs is of value in the drawing of vertical sections; it also provides a method of obtaining angles of dip of particular horizons.

Use of Stereometer

A stereometer (Plate IIA) provides a convenient method of measuring parallax differences. It carries two glass plates, which are inscribed with a very small dot. The separation of the plates along the bar can be altered by both coarse and fine adjustments; the fine adjustment is controlled by a micrometer giving direct readings to hundredths of a millimetre and enabling thousandths of a millimetre to be estimated.

The procedure for using a stereometer to measure the parallax difference of two points of different elevation is as follows:

(1) Set the micrometer reading to mid-run.

(2) Put the stereometer over the correctly set-up photographs, and adjust the separation of the plates, by means of the coarse adjustment, until the dots approximately coincide with the points of lower elevation on the two photographs (e.g. points *b* in Fig. 8).

(3) Study the stereomodel with a mirror stereoscope having a high-power binocular attachment; the dots should fuse, and apparently form a single dot either above or below the point of lower elevation on the stereomodel. If the dots will not fuse when the photographs are in fusion, alter the separation of the glass plates of the stereometer slowly, using the coarse adjustment. The dots will then suddenly fuse to form a single dot above the stereomodel. Clamp the coarse adjustment. If the micrometer is now altered, it will be found that the fused dot can be floated up and down relative to the stereomodel. Using the micrometer, float the dot downwards until it appears to rest on the stereomodel at the same elevation as that of the lower point. The distance separating the two dots on the stereometer is now identical with that of the two images of the lower point (e.g. d_1 in Fig. 8). Read the micrometer as accurately as possible and estimate the third decimal place.

(4) Using the micrometer, float the dot upwards until it appears to rest on the stereomodel at the same elevation as that of the higher point. The distance separating the two dots is now identical with that of the two images of the upper point (e.g. d_2 in Fig. 8). Read the micrometer as in (3) above. The difference between the two micrometer readings (3) and (4) is the difference in parallax of the two points.

In practice, several micrometer readings are taken for both the lower and upper points and then averaged; this increases the accuracy obtainable and obviates the possibility of a gross error in reading the micrometer.

Height Difference Formulae

It is now necessary to obtain formulae connecting differences in parallax with differences in height (Allum, 1969).

Let P_x = Parallax of point X.

P_y = Parallax of point Y.

p = Difference in parallax = $P_x - P_y$.

f = Focal length of lens (usually recorded on photograph).

B = Length of air base in the air.

W = Mean of the distances between the principal point and the transferred principal point, measured on both photographs of the stereopair.

H = Height of aircraft above sea level.

h = Height of datum level above sea level.

h_x = Height of point X above sea level.

h_y = Height of point Y above sea level.

S = Scale.

Ideally, the datum level should be intermediate between points X and Y; then, if the elevation differences between h, h_x, and h_y are small compared with H, the scale between points X and Y may be regarded as $\frac{f}{(H - h)}$;

$$\text{i.e.}\quad S = \frac{f}{(H - h)}. \qquad (1)$$

In practice, the datum level frequently has to be chosen arbitrarily from what topographical information is available.

In Fig. 23, X is a point on the ground, and XT_1 and XT_2 represent rays to two successive camera stations T_1 and T_2, which are separated in the air by a distance B. KL and MN represent the positives of two aerial photographs taken with a camera of focal length f. The two rays XT_1 and XT_2 intersect the photographs at Q and R respectively, so that Q and R represent the images of point X on the photographs. The distances of Q and R from the principal points O_1 and O_2 are L_1 and L_2 respectively.

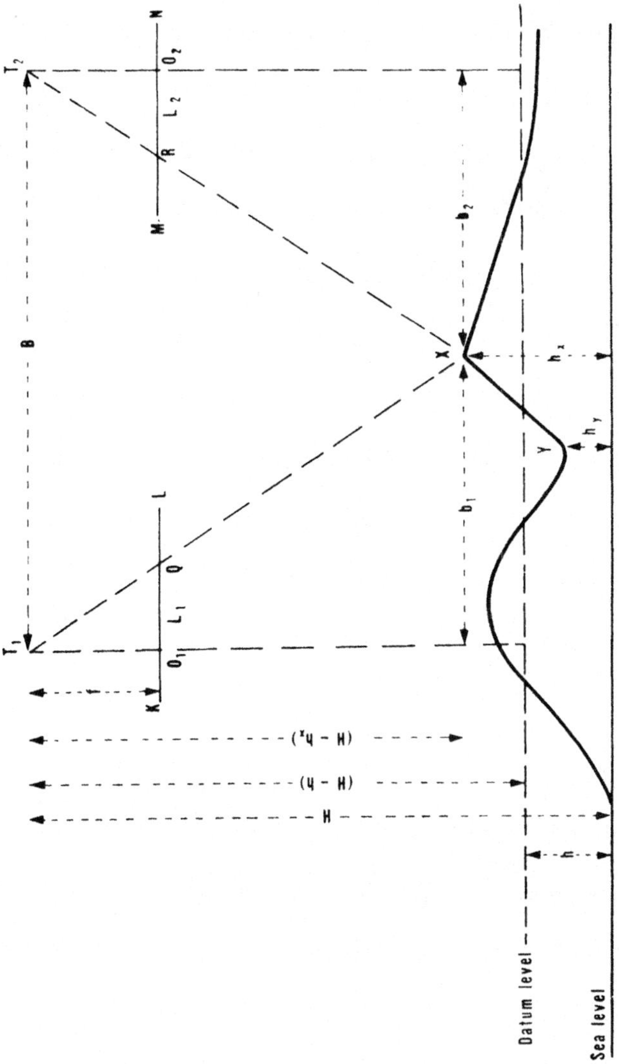
T1
T2
B
M
R
L2
O2
N
L
Q
L1
O1
K
f
X
Y
b1
b2
hx
hy
h
H
(H − h)
(H − hx)
Datum level
Sea level

FIG. 23.

From similar triangles:

$$\frac{f}{L_1} = \frac{(H - h_x)}{b_1} \text{ and } \frac{f}{L_2} = \frac{(H - h_x)}{b_2},$$

$$\therefore L_1 = \frac{fb_1}{(H - h_x)} \tag{2}$$

$$\text{and } L_2 = \frac{fb_2}{(H - h_x)}. \tag{3}$$

The parallax of point X is $L_1 + L_2$ (see Chapter 2).

$$\therefore P_x = L_1 + L_2.$$

Substituting the values for L_1 and L_2 from (2) and (3) gives:

$$P_x = \frac{fb_1}{(H - h_x)} + \frac{fb_2}{(H - h_x)}$$

$$= \frac{f(b_1 + b_2)}{(H - h_x)}.$$

But from Fig. 23, $b_1 + b_2 = B$,

$$\therefore P_x = \frac{fB}{(H - h_x)}. \tag{4}$$

Similarly,

$$P_y = \frac{fB}{(H - h_y)}. \tag{5}$$

It can be seen from equations (4) and (5) that for a given photograph format (i.e. when the focal length of camera, height of aircraft above sea level, and the air base remain the same), all points of the same elevation have the same parallax, and points with a higher elevation have a greater parallax.

But

$$p = P_x - P_y.$$

Substituting the values for P_x and P_y from (4) and (5) gives:

$$p = \frac{fB}{(H - h_x)} - \frac{fB}{(H - h_y)},$$

$$\therefore p = \frac{fB\,(h_x - h_y)}{(H - h_x)\,(H - h_y)}. \tag{6}$$

If h is intermediate between h_x and h_y, and if the difference between h_x and h_y is small compared to H, then:

$$(H - h_x)(H - h_y) = (H - h)^2 \tag{7}$$

approximately.

Substituting $(H - h)^2$ for $(H - h_x)(H - h_y)$ in (6) gives:

$$(h_x - h_y) = \frac{p\,(H - h)^2}{fB}. \tag{8}$$

But B is the air base as measured in the air; it is equal to the mean distance between the principal points and the transferred principal points measured on the photographs and divided by the scale; i.e. $B = \dfrac{W}{S}$.

But
$$S = \frac{f}{(H - h)}$$

(see equation (1)),

$$\therefore B = \frac{W(H - h)}{f}.$$

Substituting this value for B in (8) gives:

$$(\mathbf{h_x} - \mathbf{h_y}) = \frac{(\mathbf{H} - \mathbf{h})\mathbf{p}}{\mathbf{W}}. \tag{9}$$

Equation (9) **enables the height difference** $(h_x - h_y)$, **between points** X **and** Y, **to be found, providing** $(H - h)$ **is known.** The method is as follows:

(1) Obtain W by measuring the distance between the principal point and transferred principal point, on both photographs of the stereopair, and taking the mean. (A millimetre scale is suitable, and fractions of a millimetre should be estimated.)
(2) Measure p with a stereometer.
(3) Decide upon a value for h from local topographical knowledge or from existing maps.
    H will be recorded on the photographs.
    Calculate the value of $(H - h)$.
(4) Substitute these values of W, p, $(H - h)$ in equation (9) to give the height difference between points X and Y. W and p must be measured in the same units (e.g. millimetres) and then $(h_x - h_y)$ will be in the same units as $(H - h)$. Thus the same formula can be used to find $(h_x - h_y)$ in either metres or feet depending on the units of $(H - h)$.

The obvious weakness of this method is that h has to be arbitrarily chosen. If a reliable planimetric map of the area already exists, it is possible to avoid using h. All that is necessary is to obtain a mean value for the scale S by measuring the distance between two clearly defined points both on the map and on the photographs. (If the map indicates the photographs principal points, these should be used.)

Then
$$S = \frac{f}{(H - h)}, \tag{1}$$

$$\therefore (H - h) = \frac{f}{S}.$$

Substituting for $(H - h)$ in (9) gives:

$$(\mathbf{h_x} - \mathbf{h_y}) = \frac{\mathbf{fp}}{\mathbf{WS}}. \tag{10}$$

Equation (10) is the formula to use to determine height differences, if it is possible to obtain a reliable value for S, as described above. No estimation is required in its use, because f is recorded on the photographs, and W, S, and p are measured. W is measured as described when formula (9) is used (see stage (1) of above method).

If p and W are in the same units, e.g. millimetres, then $(h_x - h_y)$ will be in the same units as f. Thus, if $(h_x - h_y)$ is required in feet, and f is given as 6 in., the value of f to insert into formula (10) is 0·5 ft.

If there is a point of known elevation on the common overlap, additional spot heights may be added using the above method. The work is not so laborious as it might appear, because, for a given overlap, p is the only variable on the right-hand sides of equations (9) and (10). These formulae are equivalent to $h_x - h_y = Kp$, where K is a constant for the overlap, which needs be calculated only once. When K is known, a slide rule enables the height difference between the known point and the new point to be found very quickly.

Formulae (9) and (10) become progressively less accurate as conditions depart from those specified immediately after equation (6). It is possible to find alternative formulae that are more suitable for these new conditions (Allum, 1969). For instance, if $(h_x - h_y)$ is more than about 3 per cent of the flying height, the following formulae are to be preferred:

$$h_x - h_y = \frac{(H - h)\,p}{W + p}, \qquad \text{instead of (9),}$$

$$h_x - h_y = \frac{fp}{(W + p)S}. \qquad \text{instead of (10).}$$

Ray (1960) records formulae suitable for use in areas of high relief, when the height of the aeroplane above the upper or lower point is known.

Dip Formula

We can use our height difference formula to provide us with a formula for measuring dip.

Let α = Angle of dip from X to Y (see Fig. 24).
L = Horizontal distance between points X and Y on the ground.
C = Mean of the distance between points X and Y, measured on both photographs of the stereopair.

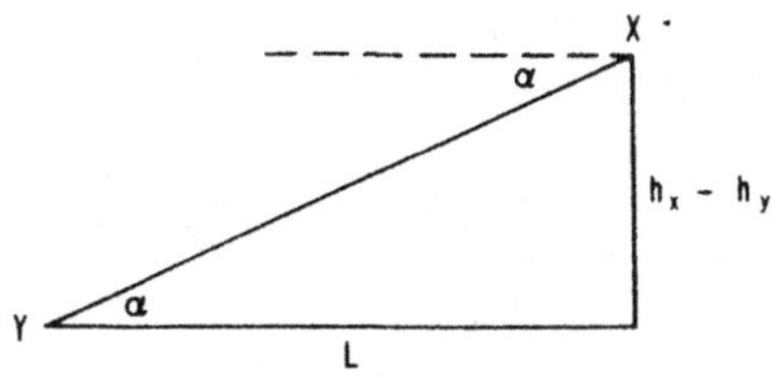

FIG. 24.

From Fig. 24 it can be seen that:

$$\tan \alpha = \frac{h_x - h_y}{L}. \qquad (11)$$

But $$L = \frac{C}{S},$$

$$\therefore L = \frac{C(H-h)}{f} \quad \text{because } S = \frac{f}{(H-h)}.$$

But $$(h_x - h_y) = \frac{(H-h)p}{W}. \qquad \text{from (9)}$$

Substituting for L and $(h_x - h_y)$ in (11) gives:

$$\mathbf{\tan \alpha = \frac{fp}{CW}}. \qquad (12)$$

Equation (12) enables the angle of depression between two points to be measured; thus if, on a given sedimentary horizon, two points (X and Y) are chosen lying on a line of true dip, the dip of the sedimentary horizon can be measured.

The method is:

(1) Obtain W as described for measurement of height difference.

(2) Obtain C by measuring the distance between points X and Y on both photographs of the stereopair, and taking the mean.

(3) Measure p with a stereometer.

(4) Note f, which should be recorded on the photographs. Substitute the values obtained, in equation (12), to give the value of tan α. Look up α in tangent tables. All values should be in the same units, e.g. millimetres or inches (1 in. = 2·54 cm).

Formula (12) is only approximately correct. It includes the approximations in the formula for the height difference, and it also ignores the different amount of relief displacement of points X and Y, which itself depends upon the height difference between points X and Y.

CONTOURS AND FORM LINES

Photogrammetrists, with adequate ground control, and using highly sophisticated instruments, can contour an area with an accuracy comparable with that of the classical field methods. The production of such contours is beyond the scope of this book.

Uncontrolled contour lines, however, which are known as **form lines,** can be produced by the geologist working by himself without special equipment. These can be of value in special areas, where for economic or other reasons, it is essential for the relief to be indicated.

Form lines may be drawn merely by putting a stereopair under a mirror stereoscope and annotating one photograph with lines that appear to lie on the stereomodel at constant elevation.

If there is a point of known elevation on the common overlap, additional spot heights may be added, using the method described in the section "Measurement of Height Differences" and more reliable form lines may be drawn by interpolating between them.

If there are two points of known, different elevation, on the common overlap, much time in determining the additional spot heights can be saved by taking parallax readings for both points with a stereometer and plotting a graph of the parallax reading against elevation. The elevation equivalents, of all intermediate parallax readings, may then be read off the graph. (It does not matter what the original coarse adjustment of the stereometer is, so long as it is such that it does not have to be altered throughout the operation.)

Instruments may be obtained that consist of a combination of stereometer and drawing attachment. These can be used for drawing form lines. If there are two points of known elevation on the common overlap, the procedure is as follows. First, the stereometer is used to measure the parallax difference between the two points. This figure is then used to calculate by simple proportion the parallax difference equivalent to the height of the first form line above the lower point. The stereometer micrometer is altered by the amount of this parallax difference. The stereometer is moved until the fused dot appears to lie on the stereomodel. The drawing attachment is then brought into operation and the stereometer is moved so that the fused dot appears to lie always on the stereomodel; the drawing attachment thus traces out the form line. The process is repeated for all the different form-line intervals.

It is possible to obtain structural contours, which are of particular interest to oil geologists, from aerial photographs. Readers interested in this subject are recommended to consult the papers by Desjardins (1950a, b).

References

ADMIRALTY (1948) *Admiralty Manual of Hydrographic Surveying,* H.M.S.O., London.

ALLUM, J. A. E. (1961) *Photogeological interpretation of areas of regional metamorphism,* Transactions of the Institution of Mining and Metallurgy, London, **70**, 1960–1, Pt. 9 (Bulletin of the Institution of Mining and Metallurgy, London, No. 655, June).

ALLUM, J. A. E. (1969) *Proofs and Uses of Some Formulae for the Measurement of Height Differences and Dips on Aerial Photographs by Elementary (Stereometer, Ruler, and Pencil) Methods,* Transactions of the Institution of Mining and Metallurgy, London (Section B: Applied Earth Science) 78, B108-B113.

ALLUM, J. A. E. (1970) *Consideration of the Relative Values of True and Infrared Colour Aerial Photography for Geological Purposes,* Transactions of the Institution of Mining and Metallurgy, London (Section B: Applied Earth Science) **79**, B76-B87.

AMERICAN SOCIETY OF PHOTOGRAMMETRY (1960) *Manual of Photographic Interpretation,* American Society of Photogrammetry, U.S.A.

AMERICAN SOCIETY OF PHOTOGRAMMETRY (1966) *Manual of Photogrammetry* (Third Edition) 2 vols. American Society of Photogrammetry, U.S.A.

AMERICAN SOCIETY OF PHOTOGRAMMETRY (1968) *Manual of Colour Aerial Photography,* American Society of Photogrammetry, U.S.A.

ANON. (1963) *Notes on Photogeology,* Economic Commission for Asia and the Far East: Committee on Industry and Natural Resources, I & NR/SGWP (5)/34.

AVERY, T. E. (1977) *Interpretation of Aerial Photographs,* 3rd Edition. Burgess Publishing Co., U.S.A.

CORTEN, F. L. (1962) *Rational planning and execution of aerial photography,* International Archives of Photogrammetry, vol. XIV (Transactions of the Symposium on Photo Interpretation, Delft, The Netherlands).

DEPARTMENT OF THE ARMY (1955) *A Guide to the Compilation and Revision of Maps,* TM5-240, Department of the Army, Washington, D.C.

DESJARDINS, L. (1950a) *Structural contouring for the photogeologist,* Photogrammetric Engineering, 16, No. 5, December.

DESJARDINS, L. (1950b) *Techniques in photogeology,* Bulletin of the American Association of Petroleum Geologists, **34**, No. 12, December.

FISCHER, W. A. (1958) *Colour aerial photography in photogeologic interpretation,* Photogrammetric Engineering, **24**, No. 4, September.

GREENWOOD, J. E. G. W. (1962) *Rock weathering in relation to the interpretation of igneous and metamorphic rocks in arid regions,* International Archives of Photogrammetry, vol. XIV (Transactions of the Symposium on Photo Interpretation, Delft, The Netherlands).

HARKER, A. (1950) *Metamorphism,* Methuen, London.

KILFORD, W. K. (1963) *Elementary Air Survey,* Pitman, London.

LAHEE, F. H. (1941) *Field Geology,* McGraw-Hill, New York.

LUEDER, D. R. (1959) *Aerial Photographic Interpretation,* McGraw-Hill, New York.

MILLER, V. C. (1961) *Photogeology,* McGraw-Hill, New York.

NEWTON, A. R. (1971) *The Uses of Photogeology: A Review,* Transactions of the Geological Society of South Africa, Vol. LXXIV, Part III, pp. 149–171.

NORMAN, J. W. *et al.* (1971) *Discussion of Paper* (Allum 1970). Transactions of the Institution of Mining and Metallurgy, London (Section B: Applied Earth Science) **80**, B122-B132.

VERSTAPPEN, H. Th. (1977) *Remote Sensing in Geomorphology,* Elsevier Scientific Publishing Co., The Netherlands.

Recommended Reading

CHAPTER 1

ALLUM, J. A. E. (1970) *Consideration of the Relative Values of True and Infrared Colour Aerial Photography for Geological Purposes*. Transactions of the Institution of Mining and Metallurgy, London (Section B: Applied Earth Science) **79**, B76-B87.

AMERICAN SOCIETY OF PHOTOGRAMMETRY (1968) *Manual of Colour Aerial Photography*, American Society of Photogrammetry, U.S.A.

AVERY, T. E. (1977) *Interpretation of Aerial Photographs*, 3rd Edition. Burgess Publishing Co., U.S.A.

BROCK, G. C. (1967) *The Physical Aspects of Aerial Photography*. Dover Publications, Inc., New York.

MEKEL, J. F. M. (1974) *A Short History of Photogeology*. Information of the International Institute for Aerial Survey and Earth Sciences (ITC), Enschede, The Netherlands, Number 30.

NEWTON, A. R. (1971) *The Uses of Photogeology: A Review*, Transactions of the Geological Society of South Africa, Vol. LXXIV, Part III, pp. 149–171.

NORMAN, J. W. *et al.* (1971) *Discussion of Paper* (Allum 1970). Transactions of the Institution of Mining and Metallurgy, London (Section B: Applied Earth Science) **80**, B122-B132.

WORSFOLD, R. D. (1976) *Colour Compensation Filters with Infrared Film*, Photogrammetric Engineering and Remote Sensing, Vol. XLII, No. 11, pp. 1385–1398.

CHAPTERS 4, 5 and 6

AMERICAN SOCIETY OF PHOTOGRAMMETRY (1960) *Manual of Photographic Interpretation*, American Society of Photogrammetry, U.S.A.

LUEDER, D. R. (1959) *Aerial Photographic Interpretation*, McGraw-Hill, New York.

MEKEL, J. F. M. (1970) *The Use of Aerial Photographs in Geological Mapping*, ITC Textbook of Photo-Interpretation, Volume VIII (1). International Institute for Aerial Survey and Earth Sciences (ITC), The Netherlands.

MILLER, V. C. (1961) *Photogeology*, McGraw-Hill, New York.

NEWTON, A. R. *et al.* (1973) *Photogeology in a Granitic Gneiss Terrain: Namaqualand, South Africa*, Special Publication Geological Society of South Africa, 3, pp. 277–287.

NORMAN, J. W. (1976) *Photogeological Fracture Trace Analysis as a Subsurface Exploration Technique*, Transactions of the Institution of Mining and Metallurgy, London (Section B: Applied Earth Science) **85**, B52-B62.

RAY, R. G. (1961) *Aerial Photographs in Geologic Interpretation and Mapping*, U.S. Geological Survey, Professional Paper 373.

VERSTAPPEN, H. Th. (1977) *Remote Sensing in Geomorphology*, Elsevier Scientific Publishing Co., The Netherlands.

CHAPTERS 7, 8 and 9

LATTMAN, L. H. *et al.* (1965) *Aerial Photographs in Field Geology*, Holt, Rinehart and Winston.

CHAPTER 10

ALLUM, J. A. E. (1969) *Proofs and Uses of Some Formulae for the Measurement of Height Differences and Dips on Aerial Photographs by Elementary (Stereometer, Ruler, and Pencil) Methods*, Transactions of the Institution of Mining and Metallurgy, London (Section B: Applied Earth Science) **78**, B108-B113.

AMERICAN SOCIETY OF PHOTOGRAMMETRY (1966) *Manual of Photogrammetry* (Third Edition) 2 vols. American Society of Photogrammetry, U.S.A.

Index

(More important references are indicated by bold type)

Printed in Dunstable, United Kingdom

91221454R00080